Lecture Notes of the Institute for Computer Sciences, Social Informatics and Telecommunications Engineering 175

More information about this series at http://www.springer.com/series/8197

Jia Hu · Victor C.M. Leung
Kun Yang · Yan Zhang
Jianliang Gao · Shusen Yang (Eds.)

Smart Grid Inspired Future Technologies

First International Conference, SmartGIFT 2016
Liverpool, UK, May 19–20, 2016
Revised Selected Papers

 Springer

Editors
Jia Hu
Department of Computer Science
University of Exeter
Exeter
UK

Victor C.M. Leung
Department of Electrical and Computer
  Engineering
The University of British Columbia
Vancouver, BC
Canada

Kun Yang
University of Essex
Essex
UK

Yan Zhang
Simula Research Laboratory
Fornebu
Norway

Jianliang Gao
Imperial College
London
UK

Shusen Yang
The Institute of Information and System
  Science
Xi'an Jiaotong University
Xi'an
China

ISSN 1867-8211   ISSN 1867-822X (electronic)
Lecture Notes of the Institute for Computer Sciences, Social Informatics
and Telecommunications Engineering
ISBN 978-3-319-47728-2   ISBN 978-3-319-47729-9 (eBook)
DOI 10.1007/978-3-319-47729-9

Library of Congress Control Number: 2016957484

Printed on acid-free paper

This Springer imprint is published by Springer Nature
The registered company is Springer International Publishing AG
The registered company address is: Gewerbestrasse 11, 6330 Cham, Switzerland

Preface

We are very pleased to introduce the proceedings of the First EAI International Conference on Smart Grid Inspired Future Technologies (SmartGIFT 2016). This was the first SmartGIFT conference, aiming to create a forum for researches, developers, and practitioners from both academia and industry to publish their key results and to disseminate state-of-the-art concepts and techniques in all aspects of smart grids. The 37 scientific participants had many fruitful discussions and exchanges that contributed to the success of the conference. Participants from 12 countries made the conference truly international in scope.

SmartGIFT covers many technical topics including high-level ideology/methodology, concrete smart grid inspired data sensing/processing/networking technologies, smart grid system architecture, quality-of-service (QoS), energy efficiency, security in smart grid systems, management of smart grid systems, service engineering and algorithm design, and real-world deployment experiences, etc. The peer-reviewed technical program of SmartGIFT 2016 consisted of 25 full papers in oral presentation sessions from the main conference track, the Workshop on Security, Flexibility and Demand Management in Smart Low-Voltage Grids, the Workshop on Systems and Technology for Energy Management in Communication Networks, and the special track on Communications and Management of Smart Grid Enabled Cities. Aside from high-quality paper presentations, the technical program also featured a keynote speech from world-renowned scientist, Prof. Vincent Wong from the University of British Columbia, Canada, and a panel discussion on "ICT-Driven Innovation for the Energy Field" organized by EIT Digital.

Coordination with the steering chairs, Imrich Chlamtac and Victor C.M. Leung, and the general chairs, Victor C.M. Leung and Kun Yang, was essential for the success of the conference. We sincerely appreciate their constant support and guidance. It was a great pleasure to work with such an excellent team. We are also indebted to the workshop chairs, Shusen Yang and Ming Xia, for their excellent work in organizing the attractive workshops, and to the publication chair, Jianliang Gao, for his hard work in preparing these proceedings. We would like to express our appreciation to all the members of the Organizing Committee and Program Committee for their efforts and support. We are grateful to all the authors who submitted their papers to the SmartGIFT 2016 conference and workshops.

Given the rapidity with which science and technology is advancing in all the fields covered by SmartGIFT, we expect that future SmartGIFT conferences will be as stimulating as this very first one was, as indicated by the contributions presented in this volume.

October 2016

Jia Hu
Yan Zhang

Organization

Steering Committee

Imrich Chlamtac   Create-Net and University of Trento, Italy
Victor C.M. Leung   The University of British Columbia, Canada

Organizing Committee

General Chairs

Victor C.M. Leung   The University of British Columbia, Canada
Kun Yang   University of Essex, UK

Technical Program Committee Chairs

Jia Hu   University of Exeter, UK
Yan Zhang   Simula Research Laboratory, Norway

Publicity and Social Media Chair

Lei Liu   Shandong University, China

Publication Chair

Jianliang Gao   Imperial College, UK

Workshops Chairs

Shusen Yang   Xi'an Jiaotong University, China
Ming Xia   Ericsson Research USA

Sponsorship and Exhibits Chair

Mark Barrett-Baxendale   Liverpool Hope University, UK

Local Chair

Jia Hu   University of Exeter, UK

Web Chair

Stewart Blakeway   Liverpool Hope University, UK

Conference Manager

Barbara Fertalova   EAI

Technical Program Committee

Farrokh Aminifar	University of Tehran, Iran
Islam S. Bayram Hamad bin	Khalifa University, Qatar
Subhonmesh Bose	Cornell University, USA
Shengrong Bu	University of Glasgow, UK
Luisa Caeiro	EST/IPS, Portugal
Hong-An Cao	ETH Zurich, Switzerland
Alvaro Cardenas	University of Texas, Dallas, USA
Stephan Cejka	Siemens AG Österreich, Austria
Chen Chen	Argonne National Laboratory, USA
Lijun Chen	University of Colorado at Boulder, USA
C.G. Deepak	Lancaster University, UK
Yingfei Dong	University of Hawaii, USA
Felicita Di Giandomenico	Italian National Research Council, ISTI, Italy
Lorenza Giupponi	CTTC, Barcelona, Spain
Ali Haghighi	INRS-EMT, Canada
Syed Faraz Hasan	Massey University, New Zealand
Miao He	Texas Tech University, USA
Hossein Akhavan Hejazi	University of California at Riverside, USA
Melike Erol Kantarci	Clarkson University, USA
Amin Khodaei	University of Denver, USA
Younghun Kim	IBM T.J. Watson Research Center, USA
David Koilpillai	IIT Madras, India
Peng-Yong Kong	Khalifa University of Science, Technology and Research, UAE
Albert YS Lam	The University of Hong Kong, Hong Kong, SAR China
Long Le	INRS, University of Quebec, Canada
Tuan Le	Middlesex University, UK
Shunbo Lei	The University of Hong Kong, Hong Kong, SAR China
Qinghua Li	University of Arkansas, USA
Hui Lin	Fujian Normal University, China
Rongxing Lu	Nanyang Technological University, Singapore
Chunbo Luo	University of Exeter, UK
Lotfi Mhamdi	Leeds University, UK
Hamed Mohsenian-Rad	University of California at Riverside, USA

Keivan Navaie	Lancaster University, UK
Duy Ng	University of Newcastle, Australia
Hieu Trung Nguyen	INRS, University of Quebec, Canada
Qian Ni	Lancaster University, UK
Dusit Niyato	Nanyang Technological University, Singapore
Moslem Noori	University of Alberta, Canada
Jorge Ortiz	IBM T.J. Watson Research Center, USA
Zhibo Pang	BB AB Corporate Research, Sweden
Sadigheh Parsaifard	ITRC, Iran
Prashant Pillai	University of Bradford, UK
Tony Q.S. Quek	Singapore University of Technology and Design, Singapore
Haile-Selassie Rajamani	University of Bradford, UK
Harold Rene Chamorro Vera	KTH, Sweden
Emanuele Lindo Secco	Liverpool Hope University, UK
Boon-Chong Seet	Auckland University of Technology, New Zealand
Babak Seyfe	Shahed University, Iran
Vahid Shah-Mansouri	University of Tehran, Iran
Zhengguo Sheng	University of Sussex, UK
Hongjian Sun	Durham University, UK
Hissam Tawfik	Leeds Beckett University, UK
Ye Tian	Beijing University of Posts and Telecommunications, China
Nghi Tran	The University of Akron, USA
Khan-Ferdous Wahid	Airbus, Germany
Xianbin Wang	Western University, Canada
Yulei Wu	University of Exeter, UK
Yong Xiao	University of Houston, USA
Junfeng Xu	China Information Technology Security Evaluation Center, China
Yunjian Xu	Singapore University of Technology and Design, Singapore
Po Yang	Liverpool John Moores University, UK
Qiang Yang	Zhejiang University, China
F. Richard Yu	Carleton University, Canada
Li Zhang	University of Leeds, UK
Haijun Zhang	The University of British Columbia, Canada
Lu Zhuo	University of Memphis, USA

MASTERING Workshop

Organization (Workshop Co-chairs)

Monjur Mourshed	Cardiff University, UK
Maryse Anbar	Engie, France
Thomas Messervey	R2M Solution, Italy
Meritxell Vinyals	French Alternative Energies and Atomic Energy Commission (CEA), France

Program Committee

Zia Lennard	Juan Espeche, R2M Solution, Italy
Sylvain Robert	Hassan Sleiman, French Alternative Energies and Atomic Energy Commission (CEA), France
Mario Sisinni	Robert-Erskine Murray, SMS plc., UK
Yann Pankow	Laborelec, France
Ivan Grimaldi	Telecom Italia, Italy
Laura Rodriguez Martin	Airbus DS Cybersecurity, France
Hisain Elshaafi	Telecommunications Software and Systems Group (TSSG)/Waterford Institute of Technology (WIT), Ireland
Michael Dibley	Cardiff University, UK

Special Session on Communications and Management of Smart Grid-Enabled Cities

Session Organizers

Michael Chai	Queen Mary University of London, UK
Yue Chen	Queen Mary University of London, UK

STEMCOM Workshop

General Chair

Muhammad Ali Imran	University of Surrey, UK

General Co-chair

Mounir Ghogho	University of Leeds, UK

Technical Program Chairs

Syed Ali Raza Zaidi   University of Leeds, UK
Muhammad Zeeshan   Texas A&M University at Qatar, Doha
  Shakir
Mischa Dohler   King's College London, UK
Ranga Rao Venkatesha   Delft University of Technology, The Netherlands
  Prasad

Contents

MASTERING Workshop

Special Session Track

Invited Papers Track

Main Track

Mobility Incorporated Vehicle-to-Grid (V2G) Optimization for Uniform Utilization in Smart Grid Based Power Distribution Network

Muhammad A. Hussain[✉] and Myung J. Lee

Department of Electrical Engineering,
City University of New York, City College, New York, USA
mhussai09@citymail.cuny.edu, mlee@ccny.cuny.edu

Abstract. V2G technology in smart grid architecture enables the bidirectional flow of electric energy where a Plug-in Electric Vehicle (PEV) can also discharge energy to the grid from its battery. Thus when a good number of PEVs are available, for instance, in a big parking lot enabled with Electric Vehicle Supply Equipment (EVSE), they form a large distributed energy storage system. A controller which can communicate with such EVSE enabled parking lots can optimally control charging/discharging schedules of each PEV to minimize the peak load demand of the distribution grid. While minimizing the peak demand, our optimization strategy also considers improved distribution of PEV loads throughout the distribution network to minimize the impact in any particular feeder or transformer utilizing mobility information by Vehicular Ad hoc Network (VANET) communications. This novel algorithm of uniform utilization can substantially reduce the need of expensive infrastructure upgrade of power distribution network.

Keywords: V2G optimization · VANET · Smart grid · Plug-in electric vehicle · EVSE parking lot · Utilization · Mobility

1 Introduction

According to the U.S. Department of Energy (DOE), transportation is the largest emitter of carbon dioxide in the United States, accounting for roughly one third of all CO_2 emissions. In this context Plug-in Electric Vehicles (PEV) can play a major role of minimizing greenhouse gas emission. The PEV market is growing rapidly and the penetration is forecasted as high as 26.9 % by 2023 and 72.7 % by 2045 [1]. But electric energy requirement of PEVs is quite high. A single PEV can consume as much energy as an air-conditioned home—potentially doubling peak residential electricity demand. Uncontrolled use of charging may increase risk of overburdening the existing distribution network grid such as undesirable peaks and unwanted loads [2]. The load demand varies with time of day, week etc. due to weather, human activity and some random factors such as time of using different household and commercial electrical appliances. The most significant challenge of the power distribution infrastructure is peak load demand. The designing of the electricity infrastructure capacity is based upon peak load demand as it needs to deliver such amount when called upon. To cope

© ICST Institute for Computer Sciences, Social Informatics and Telecommunications Engineering 2017
J. Hu et al. (Eds.): SmartGIFT 2016, LNICST 175, pp. 3–15, 2017.
DOI: 10.1007/978-3-319-47729-9_1

with peak demand increase, electric utility companies, even after the intensive energy efficiency programs, often need to undergo expensive capacity upgrading by installing new transformers, reinforcing distribution feeders, transformers, etc., that are projected to operate beyond 100 % of their ratings [9]. A critical bottleneck of deeper penetration of PEVs into the automotive market is that if large number of PEVs draw current from the grid during the peak load situation, there can be significant amount of increase in the peak demand triggering capacity enhancement of existing electric infrastructure. This threat to utility grid also draws attention to Independent System Operators of the electric power system and Regional Transmission Organizations (ISO/RTO) [3]. To overcome these challenges, smart grid technologies will be the primary means to manage electric vehicle charging by a proper communication architecture between PEVs and the grid. Extensive research is underway to develop standards such as a framework released by National Institute of Standards and Technology for Smart Grid Standards covering interoperability standards to support plug-in electric vehicles [4].

With the advent of Vehicle to Grid (V2G) technology, PEVs can also discharge power to the grid. Although discharging of single PEV individually cannot contribute much for load balancing and peak shaving, many PEVs together can significantly improve these characteristics with V2G optimization. V2G application constitutes both bidirectional power flow which includes electricity discharging from vehicle to grid and the charging PEV with a rate control [5]. By utilizing the instant response characteristics of battery packs installed on PEVs as distributed energy storage, V2G can offer ancillary services such as demand-response, frequency regulation, and spinning reserves [6]. The optimizations for maximizing the profit by utilizing PEVs' battery resources for ancillary services are well studied. These optimization approaches, however, assumed the PEVs were already connected to the Electric Vehicle Supply Equipment (EVSE) stations. The knowledge of a PEV's mobility information such as current location, current state of charge, preferred EVSE stations, expected arrival time at destination, arrival state of charge, intended departure state of charge even before its arrival at the EVSE station can be very advantageous in the V2G optimization problem. The utilization of these priori information of a PEV's mobility status by VANET communication has just begun [7]. The research of Intelligent Transportation System (ITS) and VANET focuses on accurate and timely traffic flow information. With such VANET communications system, one can assume the optimization analysis can include priori information from the PEV user at the beginning of its journey to a destination. We propose an optimization algorithm with a scalable architecture that takes this priori information into account. In contrast to other works of optimization that maximizes the profit of PEV users, the objective of our optimization is to minimize the variation in utilization among different parts of the power distribution network along with shaving the peak demand. The rationale is that a certain part of the distribution network can be over utilized while other part is underutilized which can trigger an expensive infrastructure upgrade, e.g., substation capacity enhancement, transformer change, feeder cable changes [8]. This high cost of upgrading can overweigh any immediate benefit of merely looking at maximization of profits by V2G optimization. Therefore, this paper is presenting a feasible optimization architecture to minimize the variation of utilization to maximize the lifetime of existing electric infrastructure.

The remainder of this paper is organized as follows. Section 2 describes the related works of V2G optimization; Sect. 3 describes the system framework; Sect. 4 explains optimization approach; and Sect. 5 presents the simulation scenario, results and performance evaluation. Finally Sect. 6 concludes the paper.

2 Related Works in V2G Optimization

Many works in the literature addressed V2G optimization for ancillary services through coordinated charge/discharge scheduling among different PEVs by bidirectional power flow model between PEVs and grid [10–12]. Centralized scheduling by an aggregator is suggested with the concern of computational complexity of existing solution approaches such as linear programming. In [13], authors evaluate the large scale PEV parking lots and analyze the impact of aggregated load by several thousand PEVs into grid service under various charging scenarios. To tackle the computational challenge, some used metaheuristic approaches like Particle Swarm Optimization (PSO) [14]. Monte Carlo simulations are also used to evaluate the performance for practical systems [15]. Importantly, however, all of these works are limited to the scenario that optimization parameters such as departure time, arrival state of charge, desired final state of charge are only collected after PEVs are already parked in EVSE facility.

Very few papers on V2G optimization addressed the issue of mobility aspect of PEV utilizing VANET. In [16] authors include the VANET architecture in their V2G optimization work. They assumed that if a PEV needs charging in the middle of its route, it can communicate with the aggregator to inform its route information, charging preferences using the RSUs (road side units). With this provided information, the arrival time at a charging station can be estimated and aggregator can plan ahead about which charging station can be chosen in the PEV's trajectory to minimize the average cost of charging. The range anxiety of mobile PEVs are also considered in [17] where VANET enhanced smart grid is considered to provide an efficient coordinated charging strategy to route mobile EVs to fast-charging stations. These works, however, considered only the fast-charging requirement and range anxiety issues of the PEV users. Also, their optimizations focused only on current grid situation rather than long-term load-profile optimization of distribution network.

Regarding VANET, extensive literatures reported on the research issues such as VANET routing, medium access, traffic management, and a variety of connected vehicle applications [20]. Road side units (RSU) of VANET can collect the Basic Safety Message (BSM) message from vehicles and forward this information to Traffic Management Center (TMC) by backbone internet [26]. Survey of a variety of traffic management approaches is reported in [27]. A central entity TMC plays the role of providing route guidance, travel duration and other PEV information required for the optimization of V2G services. It should also be noted that in near future, autonomous driving is also projected to become a key component in the transportation industry [28]. Autonomous driving will filter out the randomness of human behavior and can reach closer to deterministic navigation system in the transportation network.

In contrast to these existing VANET integrated smart grid works, our work deals with the issue of distributing PEV loads among different network segments, so that the variance of utilization is minimized. The motivation of our work is from the fact that in busy commercial city areas where density of electricity consumption is very high, there are a lot of commuters who drive to workplace in the morning and leave in the evening. For example, in LA County, California, USA 72.3 % workers drive alone to their work [18]. With higher PEV market penetration, current parking lots will be converted to EVSE enabled parking facilities. Utilizing such PEV commuters, our approach gives an optimization technique for long period (from morning to evening) parking of PEV commuters in large scale EVSE enabled parking lots in a busy commercial area rather than range anxiety or fast charging requirement. Instead of looking at just current energy utilization scenario among different charging station spots, we intend to optimize the whole day load profile to provide uniform peak utilization in busy urban areas. We also proposed a simple convex optimization technique to provide fast, real time calculation and thus allowing very quick response to the user for a feasible implementation of such architecture. Our work involves the strategy of balancing the peak utilization among different network parts of distribution network by controlling the schedules of these PEVs' charging and discharging profiles with additional optimal EVSE parking lot selection for each specific PEV.

3 Scalable System Framework

Our System Framework for VANET based V2G Optimization is shown in Fig. 1. The hierarchy of the control plane consisting of Central Controller (CC), Zone Controller (ZC) and Local Controller (LC) is designed to provide the necessary scalability. In the distribution network plane, each EVSE parking lot is controlled via an LC, and connected to the distribution network through a step-down distribution transformer. Therefore, EVSEs represent electrical nodes on the power distribution network. In an

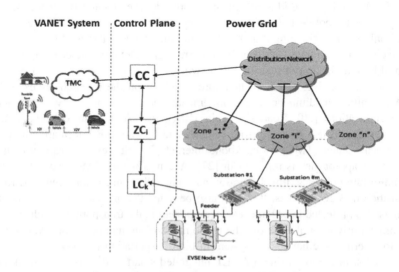

Fig. 1. System Framework of Mobility included V2G optimization

EVSE lot, PEVs are assumed to be interfaced with the distribution network through smart inverters. A smart inverter can receive an external command to draw/inject power.

Each PEV informs TMC via VANET system or internet about its preferred charging/parking zone, current status of battery, intended departure time and battery status at departure. By real time traffic analysis and management of VANET system, TMC is assumed to be capable of providing the best route navigation to the PEV user. Through VANET communications the up to date information such as required remaining travel time, any unexpected route changes of a PEV can also be tracked to reach a given destination. TMC calculates navigation routes for different preferred EVSE spots of the PEV and corresponding arrival time and arrival state of battery. TMC then reports all these information of this PEV to CC which will later be used as input parameters of optimization. CC also collects information of power distribution network by smart-grid communication [19] for necessary optimization inputs of electric distribution network such as current loads, forecasted load profile, capacity ratings at different segments, node voltages, etc. By providing necessary inputs, the CC assigns the optimization problem to the corresponding ZCs of the relevant areas. CC maintains the mapping of ZCs of relevant geographic areas. A ZC controls one or more of the EVSE lots (LCs) within the same neighborhood. ZC coordinates and aggregates optimization messages between LCs and CC. The CC collects and forwards PEV's information to the corresponding ZC, which in turn assigns the optimization problem to the corresponding LCs. Each LC will execute the optimization in parallel and send the optimization result to their ZC. After getting results from LCs, ZC can determine the optimal location and charge/discharge profile in its zone for this PEV and forwards the selection information to CC. Since each ZC only forwards the information of final selection to the CC and the optimization task resides mainly in LCs, the CC can coordinate much higher number of PEVs. By assigning the core optimization burden to the distributed LCs while ZC determines the final selection in its Zone, this architecture facilitates CC to conduct necessary inter-system communications with VANET and Power distribution network.

4   Optimization Model

For a given total energy demand, the lowest possible value of peak demand can only be the average load demand. Since $\int (Load\ Demand)dt = Energy\ Demand$, if the load demand at any particular duration is lower than the average load demand, then in some other duration it needs to be higher than the average load demand to satisfy the given total energy demand. Let us consider a base load, $f_0(t)$ which is a given forecasted demand of electricity (excluding the electricity demand by PEVs) at the node of the distribution network connected to an EVSE lot. The forecasted load curve can be collected using methods such as statistical analysis of historical data, extrapolation, time-series method, Gray theory, or artificial-neural-network methods, etc. [21–23]. We expect higher accuracy of the forecasts possible in the era of smart grid. In this paper, we assume that the forecasted load curve is given. Due to the additional energy requirement by the PEVs, if these PEVs draw their required average load demand altogether, the peak also rises by the same amount. For minimizing the peak, our optimization strategy will be focused on valley filling (charging when $f_0(t)$ is lower than the average load demand) and peak shaving (discharging when $f_0(t)$ is higher than the average load demand).

In our optimization model, initially, a PEV_i at time $T_{req,i}$ requests CC about its preferred parking area, the required final state of energy (FE_i) at departure, and the time of departure ($T_{dep,i}$). The CC will be informed about arrival time ($T_{arr,i}$) and arrival state of energy (AE_i) by TMC. The CC will assign this optimization problem to corresponding ZC which in turn forwards the assignment to relevant LCs. According to the system framework described in the previous section we can divide our V2G optimization task in three subtasks: (i) Load profile optimization at LC (ii) Optimal EVSE lot selection at ZC (iii) Final selection at CC.

Load Profile Optimization at LC: Let us consider a particular LC where the main goal is to mitigate the peak demand thus minimizing the maximum utilization of corresponding electrical element of distribution network. In the context of this optimal valley filling and peak shaving approach, we introduced a least square objective function at a particular LC on the arrival of this request from PEV_i as (1):

$$\min_{g(t)} \left(\sum_t ((f_0(t) + g(t)) - \bar{L})^2 \right) \tag{1}$$

$g(t) = \sum_{j=1}^{i} g_j(t)$, $g_j(t)$ is the load (charge/discharge) profile of any individual PEV which already requested and assigned at this LC and available after $T_{req,i}$. $g_j(t)$ is defined in either $T_{dep,j} \leq t \leq T_{req,i}$ if $T_{arr,j} \leq T_{req,i}$ or $T_{dep,j} \leq t \leq T_{arr,j}$ if $T_{arr,j} > T_{req,i}$. Let us define $T_{i,j}$ and $E_{i,j}$ as below:

$$T_{i,j} = \max(T_{req,i}, T_{arr,j}) \qquad E_{i,j} = \begin{cases} AE_j & \text{if} \quad T_{arr,j} > T_{req,i} \\ PE_j(T_{req,i}) & \text{if} \quad T_{arr,j} \leq T_{req,i} \end{cases}$$

Present battery energy, $PE_j(t) = AE_j + \sum_{T_{arr,j}}^{t} g_j(t)$ is the amount of battery energy at t

for a PEV_j which already arrived at this LC node. Average load demand ($\bar{L}$) at this node is the average of combined energy demand of base load and all requested PEVs. The optimization is performed at arrival of each request. Upon arrival of the request of PEV_i the scheduling horizon is from $T_{req,i}$ to T_{end}; i.e. t is defined as $T_{ini} \leq t \leq T_{end}$. T_{end} is considered as the maximum of $T_{dep,j}$ among $j = 1$ to i because after that none of these PEV_j will be available. If all PEVs would have arrived and depart at the same time, $\bar{L}$ will be simply $\left(\overline{f_0(t)} + \overline{g(t)} \right)$. But since PEVs will arrive and depart at different times we adapted a weighted average for $\bar{L}$ as (2):

$$\bar{L} = \frac{\sum_{j=1}^{i}\left(\left(\overline{g_j(t)} + \overline{f_j(t)}\right) \times \left(T_{dep,j} - T_{i,j}\right)\right)}{\sum_{j=1}^{i}\left(T_{dep,j} - T_{i,j}\right)} \tag{2}$$

Where, $\overline{g_j(t)} = \frac{(FE_j - E_{i,j})}{(T_{dep,j} - T_{i,j})}$ and $\overline{f_j(t)} = \frac{\sum_{T_{i,j}}^{T_{dep,j}} f_0(t)}{(T_{dep,j} - T_{i,j})}$

The constraints involved in this optimization approach include maximum current limit of a PEV, the required FE at the departure, allowed maximum and minimum threshold of battery state of energy at any instant, maximum rating of the relevant electrical element (feeder cable/transformer) at this node of electrical network. These constraints are listed in (3)–(6).

$$min\_current_j < g_j(t) < max\_current_j; \quad \forall\ PEV_j \tag{3}$$

$$\sum_{T_{i,j}}^{T_{dep,j}} g(t) = FE_j - E_{i,j}; \quad \forall\ PEV_j \tag{4}$$

$$min\_E_j < PE_j(t) < max\_E_j; \quad \forall\ PEV_j, \quad \forall t \in (T_{i,j}, T_{dep,j}) \tag{5}$$

$$f_0(t) + g(t) < L_{max}; \quad \forall t \in (T_{req,i}, T_{end}) \tag{6}$$

In other optimization approaches of V2G as discussed in related works, the common method of finding the solution is by dividing the total scheduling horizon in some number of timeslots and solving the optimal charging/discharging rate using each timeslot and each PEV as variable. Note that the number of the variables for the optimization increases both with the number of timeslots and the number of PEVs. For a good control of optimal uniform loading, the scheduling horizon should be divided in sufficiently large number of timeslots. Therefore, these techniques are not scalable for a large size EVSE lot. For a feasible mobility aware V2G application, CC must meet stringent latency constraint in responding back to the PEV user. Thus a V2G optimization approach with fewer number of variables is desirable. To address this scalability issue, we developed an innovative, simple and fast method for allocation of charge/discharge rate profile to the PEVs by fewer number of variables. Instead of finding a charging/discharging value separately at each timeslot, we seek $g_j(t)$ function with only four variables as in Eq. (7). The motivation of proposed $g_j(t)$ modelling is that the PEV should charge or discharge at a proportional rate with the difference between the base load $f_0(t)$ and $\bar{L}$; hence the variables x_{j2} and x_{j4}. By addition of offset variables x_{j1}, x_{j3} in $g_j(t)$, this model guarantees any feasible energy requirement by a PEV. The two sets of variables (x_{j1}, x_{j2}) and (x_{j3}, x_{j4}) are to apply different rates of valley filling and peak shaving for the same amount of deviation between $f_0(t)$ and $\bar{L}$ while satisfying required constraints mentioned above.

$$g_j(t) = \begin{cases} x_{j1} + x_{j2} * (\bar{L} - f_0(t)) & \text{if} \quad f_0(t) \leq \bar{L} \\ x_{j3} + x_{j4} * (\bar{L} - f_0(t)) & \text{if} \quad f_0(t) > \bar{L} \end{cases} \tag{7}$$

The optimization will solve four variables $x_{j1}, x_{j2}, x_{j3}, x_{j4}$ for each PEV which models the PEV's charge/discharge rate profile. The total scheduling period is divided into N timeslots, each slot with duration of Δt. For instance, for the granularity of $\Delta t = 1$ min and the scheduling period from 7 am to 10 pm, other approaches need to solve for 900 variables for each PEV. This modelling of $g_j(t)$ as in (7) also constructs the least square objective function (1) as a convex function. In (1), $\bar{L}$ and $f_0(t)$ are known values and $g(t) = \sum_{j=1}^{i} g_j(t)$; hence with the configuration of (7), our optimization problem i.e. objective function (1) and constraints (3) to (6) can be constructed same as the following least-square convex optimization problem.

$$\min_{x}\|cx - d\|_2^2 = \sum_{r}(c_r^T x - d_r)^2 \quad \text{Subject to } Ax \leq b$$

Where $x = [x_{11}\ x_{12}\ x_{13}\ x_{14}\ x_{21}\ x_{22}\ x_{23}\ x_{24}.\ldots\ldots\ldots x_{i1}\ x_{i2}\ x_{i3}\ x_{i4}]^T$ is the variable matrix to be solved at this LC. If this is "k"-th LC under the corresponding ZC, after finding out optimal $g(t)$, LC_k will report maximum utilization (U_k) for including this PEV$_i$ at this node to its ZC. U_k can be found as follows:

$$U_k = \frac{\max_t(f_0(t) + g(t))}{C_k} \quad \text{Where } C_k \text{ is the load} - \text{capacity (rating) at } LC_k$$

Optimal EVSE Lot Selection at ZC: The ZC will compare each LC's new peak utilization for adding this PEV$_i$ and select the one for which maximum of peak utilization among all LCs becomes minimum. Suppose in this ZC, $\{LC_1, LC_2, LC_3, \ldots, LC_k, \ldots LC_m\}$ is the set of LC nodes under this user's (PEV$_i$) preferred parking spots. Let us define, when LC_k is chosen, PU_k^1 is the highest of $\{U_1, U_2, U_3, \ldots, U_k, \ldots U_m\}$, PU_k^2 is the next highest and so on. It is to be noted that for selecting LC_k, only U_k value is updated; other values of U_1 to U_m remains as previous report. The selection matrix for choosing optimal LC is:

$$SM = \begin{pmatrix} PU_1^1 & PU_1^2 & \cdots & PU_1^m \\ & \vdots & & \vdots \\ PU_k^1 & PU_k^2 & \cdots & PU_k^m \\ & \vdots & & \vdots \\ PU_m^1 & PU_m^2 & \cdots & PU_m^m \end{pmatrix}$$

The searching for optimal LC will be conducted column-wise starting from 1st column in SM matrix because the objective is to minimize the higher utilization among all LC nodes. Suppose MP_v is the minimum of column "v" of SM matrix. Selection of the optimal LC_k is based on two criteria: (i) for some column "v" and any previous column "w" of column "v" in SM, there is exactly one row "k" for which $(PU_k^v = MP_v)$ AND $(PU_k^w = MP_w)$ (ii) if one specific "k" is not found from 1st criteria, random selection of "k" from those "r" such that *for all* $w \in \{1, 2, \ldots, m\}$, $(PU_r^w = MP_w)$. After the selection, ZC reports the optimal LC for this PEV_i in its zone to CC.

Final Selection at CC: Since the division of ZCs will be done geographically, a user's preference spots will be most likely inside the same zone. In this case, CC will determine the selection of the ZC as designated LC for this PEV_i. For any such case of a PEV_i's preferred spots fall in more than one ZC, CC selects the optimal LC from the final selections of those ZCs by same method as described for ZC in previous part. After final selection at CC, the user PEV_i is informed about the designated EVSE lot via TMC.

5   Simulation and Performance Analysis

To investigate the performance of the proposed optimization approach, we conducted simulations using MATLAB and CVX tool. We constructed our optimization scheme according to the disciplined convex programming method of CVX tool [24]. The simulation parameters that we used are listed in Table 1. We considered that 30 % users will have extra energy at arrival than they need at departure time. This is because PEV users can also charge their PEV batteries at their home after midnight at cheaper rates when the energy demand is very low. So, users who do not live far away from their working place will not lose much energy for the travel. They can get better incentive for this energy selling. For the profile of forecasted base load, we collected data of actual daily load at each hour of New York City on a certain date from NYISO (New York Independent System Operator) [25]. Then 10 different $f_0(t)$ profiles for 10 EVSE lots are made by interpolating and scaling these hourly data. The scaling is done to have an increasing peak of $f_0(t)$ from EVSE Node-1 (5 kA) to EVSE node-10 (15 kA) at 1 kA increment. We used slot duration, $\Delta t = 1$ min.

We investigated comparison between several scenarios in this simulation experiment: (i) Average Draw: each PEV goes randomly to any of this 10 nodes; each PEV

Table 1. Simulation parameters.

PEV Battery	30 kWh capacity, 30A max charge/discharge
Number of PEVs	7000
Arrival time	Normal distribution, Mean 9 am, variance 1.2
Departure time	Normal distribution, Mean 5 pm, variance 1.2
Arrival/Departure energy (% of battery capacity)	Uniform distribution
	70 % users: 10 % to 50 % arrival energy, 50 % to 90 % departure energy
	30 % users: 70 % to 90 % arrival energy, 50 % to 70 % departure energy
Energy threshold	Max: 90 %, Min: 10 % of battery capacity
EVSE lots	10 lots, each with maximum accommodation of 2000 PEVs
Load capacity rating	20 kA at each EVSE node

draws the average load that they require to acquire final energy at departure, (ii) Optimized without VANET: each PEV goes randomly to any of this 10 nodes; V2G optimization is performed at each corresponding LC node iii) Optimized with VANET: each PEV goes to the node guided by CC according to the proposed optimization model. In Fig. 2 we plotted the load profiles at four LC nodes (1,4,7,10) for each of these cases. It can be easily observed that Average Draw case strains the network most as it suffers from highest increment of peak load. Although V2G optimization is done in the case of Optimized without VANET, due to the lack of mobility information, it cannot achieve as better optimization as Optimized with VANET case. In Optimized with VANET scenario, node with lower peak loads are more utilized than the ones with higher peak loads. We could even observe the reduction of the maximum peak of base-load among all nodes (at LC node 10) even after adding all these 7000 PEVs in these 10 nodes. The maximum utilization for each case and each node is shown in Table 2. While gaining the peak shaving by V2G optimization, the optimal PEV load distribution by applying mobility information in the decision process clearly shows improved balance in utilization among different nodes.

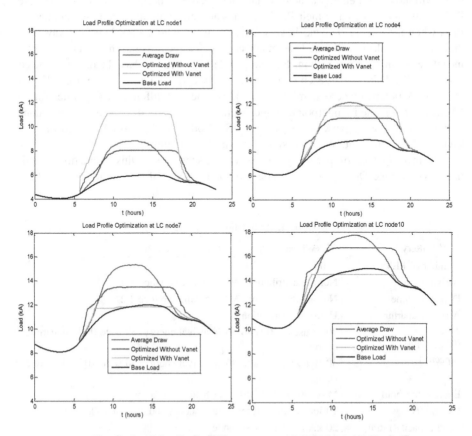

Fig. 2. Load Profile for different cases at LC nodes 1,4,7,10

Table 2. Maximum utilization (%) at each LC node for different scenarios

	Node 1	Node 2	Node 3	Node 4	Node 5	Node 6	Node 7	Node 8	Node 9	Node 10
Average draw	44.12	50.66	54.99	60.68	65.8	70.31	76.71	79.81	84.69	88.69
Optimized without VANET	40.19	43.85	48.25	54.09	58	64.37	67.44	73.76	78.09	83.61
Optimized with VANET	55.53	59.3	59.29	59.29	59.29	59.29	59.3	63.63	68.33	72.53
Base load	30	35	40	45	50	55	60	65	70	75

6 Conclusion and Future Works

This paper proposed a feasible and scalable V2G optimization architecture, and a fast optimization technique with fewer variables. By incorporating PEV mobility, our algorithm demonstrates the ability to mitigate peak demand and balance utilization and thus defer capacity upgrades for power distribution network, benefiting electric utility operators, EVSE parking lot and PEV owners. From the proposed architecture and algorithm of optimization, it can be observed that main optimization solving burden is distributed locally at LCs while ZC conducts some sorting operation which frees up CC to supervise the inter-network communication with VANET and power network. As our future work, we planned to focus on optimal incentive planning among different elements of the proposed architecture such as utility operator, PEV owners to implement such architecture. We will also work on finding the proper integration algorithm to incorporate traffic congestion parameters to the optimization function such that the V2G optimization based navigation of PEVs to different EVSE spots does not create road traffic congestion.

References

1. Duan, Z., Gutierrez, B., Wang, L.: Forecasting plug-in electric vehicle sales and the diurnal recharging load curve. IEEE Trans. Smart Grid 5(1), 527–535 (2004)
2. Sortomme, E., Hindi, M.M.: James MacPherson, S.D., Venkata, S.S.: Coordinated charging of plug-in hybrid electric vehicles to minimize distribution system losses. IEEE Trans. Smart Grid 2(1), 198–205 (2011)
3. ISO/RTO Council (IRC): Assessment of plug-in electric vehicle integration with ISO/RTO Systems, p. 31, March 2010. http://www.rmi.org/Content/Files/RTO%20Systems.pdf
4. NIST framework and roadmap for smart grid interoperability standards, release 3.0, September 2014. http://www.nist.gov/smartgrid/upload/NIST-SP-1108r3.pdf
5. Sortomme, E., El-Sharkawi, M.A.: Optimal charging strategies for unidirectional vehicle-to-grid. IEEE Trans. Smart Grid 2(1), 131–138 (2011)
6. Lin, J., Leung, K.-C., Li, V.O.K.: Optimal scheduling with vehicle-to-grid regulation service. IEEE Internet Things J. 1(6), 556–569 (2014)

7. Wang, M., Liang, H., Zhang, R., Deng, R., Shen, X.: Mobility-aware coordinated charging for electric vehicles in vanet-enhanced smart grid. IEEE J. Sel. Areas Commun. **32**(7), 1344–1360 (2014)
8. New York Independent System Operator (ISO): Power Trends 2014, Evolution of Grid. http://www.nyiso.com/public/webdocs/media_room/publications_presentations/Power_Trends/Power_Trends/ptrends_2014_final_jun2014_final.pdf
9. Con Edison Report. http://www.coned.com/messages/LICReport/Overview.pdf
10. He, Y., Venkatesh, B., Guan, L.: Optimal scheduling for charging and discharging of electric vehicles. IEEE Trans. Smart Grid **3**(3), 1095–1105 (2012)
11. Khodayar, M.E., Wu, L., Shahidehpour, M.: Hourly coordination of electric vehicle operation and volatile wind power generation in SCUC. IEEE Trans. Smart Grid **3**(3), 1271–1279 (2012)
12. Sortomme, E., El-Sharkawi, M.A.: Optimal scheduling of vehicle-to-grid energy and ancillary services. IEEE Trans. Smart grid **3**(1), 351–359 (2012)
13. Su, W., Chow, M.-Y.: Investigating a large-scale PHEV/PEV parking deck in a smart grid environment. In: North American Power Symposium (NAPS), August 2011, pp. 1–6, 4–6 (2011)
14. Su, W., Chow, M.-Y.: Performance evaluation of a PHEV parking station using particle swarm optimization. In: Proceedings of 2011 IEEE Power and Energy Society General Meeting, Detroit, Michigan, USA, 24–29 July 2011
15. Su, W., Chow, M.-Y.: Evaluation on large-scale PHEV parking deck using monte carlo simulation. In: Proceedings of the Fourth International Conference on Electric Utility Deregulation and Restructuring and Power Technologies (DRPT2011), Shandong, China, 6–9 July 2011
16. Mukherjee, J.C., Gupta, A.: A mobility aware scheduler for low cost charging of electric vehicles in smart grid. In: 2014 Sixth International Conference on Communication Systems and Networks (COMSNETS), 6–10 January 2014, pp. 1–8 (2014)
17. Wang, M., Liang, H., Zhang, R., Deng, R., Shen, X.: Mobility-aware coordinated charging for electric vehicles in VANET-enhanced smart grid. IEEE J. Sel. Areas Commun. **32**(7), 1344–1360 (2014)
18. Commuting statistics for Los Angeles County. http://factfinder2.census.gov/bkmk/table/1.0/en/ACS/11_1YR/S0801/0500000US06037
19. Li, D., Jayaweera, S.K.: Distributed smart-home decision-making in a hierarchical interactive smart grid architecture. IEEE Trans. Parallel Distrib. Syst. **26**(1), 75–84 (2015)
20. da Cunha, F.D., Boukerche, A., Villas, L., Viana, A.C., Loureiro, A.A.F.: Data communication in VANETs: a survey, challenges and applications. RR-8498, INRIA Saclay, INRIA (2014)
21. Rui, Y.: Research on the mathematical method and its application in electric load forecast. Ph.D. dissertation, Mechanical and Electrical Engineering, Central South University, Hunan, China (2006)
22. Cui, K., Li, J.R., Chen, W., Zhang, H.Y.: Research on load forecasting methods of urban power grid. Elect. Power Technol. Econom. **21**, 33–38 (2009)
23. Yao, G., Chen, Z.S., Li, X.Z.: BP network based on particle swarm optimization of short-term electric load forecasting. J. Guangdong Univ. Petrochem. Technol. **21**, 47–50 (2011)
24. Grant, M., Boyd, S.: CVX MATLAB| software for disciplined convex programming, version 1.1, September 2007. http://www.stanford.edu/~boyd/cvx/
25. NYISO (New York Independent System Operator): Hourly actual load data. http://www.energyonline.com/Data/GenericData.aspx?DataId=13&NYISO___Hourly_Actual_Load

26. Dedicated Short Range Communications (DSRC) message set dictionary, SAE Std. J2735, November 2009
27. Kamal, H., Picone, M., Amoretti, M.: A survey and taxonomy of urban traffic management: towards vehicular networks (2014). http://arxiv.org/abs/1409.4388
28. Okuda, R., Kajiwara, Y., Terashima, K.: A survey of technical trend of ADAS and autonomous driving. In: Proceedings of Technical Program - International Symposium on VLSI Technology, Systems and Application (VLSI-TSA), 28–30 April 2014, pp. 1–4 (2014)

A Cost Function Based Prioritization Method for Smart Grid Communication Network

Vahid Kouhdaragh[✉], Daniele Tarchi, Alessandro Vanelli-Coralli, and Giovanni E. Corazza

Department of Electrical, Electronic and Information Engineering, University of Bologna, Bologna, Italy {vahid.kouhdaragh2,daniele.tarchi, alessandro.vanelli,giovanni.corazza}@unibo.it

Abstract. In Smart Grids (SG) scenarios, the different nodes composing the system have to communicate to the Control Stations several type of information with different requirements. There are many communication technologies (CTs), with different Quality of Service characteristics, able to support the SG communication requirements. By focusing on wireless communications, it is possible to notice that spectrum is becoming a rare source due to its exponential increasing demand. Thus, resource allocation to support different types of SG nodes should be performed in order to maximize the resource efficiency and respect the SG requirements. Defining a cost function (CF) helps to accomplish this goal. To this aim, it is also needed to prioritize the different SG nodes based on their goals. By using the SG nodes prioritization and the CF, a priority table is defined in which the nodes and the CTs are put in order, based on their weights. The numerical results show that the proposed method allows selecting the best CT for each type of SG nodes.

Keywords: Smart grid network · Cost function · Resource allocation

1 Introduction

The conventional power grids are no more efficient and new paradigms are needed to accomplish current needs effectively: the Smart Grids (SG). There are different types of SG devices and nodes, whose number is always increasing. They report electrical power information details to the Control Station (CS) through the collectors. Demands/responses, such as dynamic consumption costs and controlling commands, are then sent back to the SG devices. Each SG cluster of nodes can have different communication characteristics, even all the type of nodes usually generate low data rate traffic. Such data are usually collected by the aggregators and transferred to the CS by using the communication technologies (CT).

Among several alternatives, the wireless CTs are considered useful for Smart Grid Communication Network, SGCN, due to several advantages [1, 2]. However, due to the SG nodes requirements, designing a SGCN based on wireless technologies becomes an important issue [1]. Among others, due to the increasing number of deployed SG nodes,

© ICST Institute for Computer Sciences, Social Informatics and Telecommunications Engineering 2017
J. Hu et al. (Eds.): SmartGIFT 2016, LNICST 175, pp. 16–24, 2017.
DOI: 10.1007/978-3-319-47729-9_2

the increased demand of resources and the requirements in terms of response latency are becoming two critical issues [1, 3]. Indeed recently, spectrum scarcity is gaining an increased interest in the research world, in particular when applied to machine type communications (MTC), where SGCN can be considered as a specific type of MTC [3]. Hence, it is even more important to find proper solutions for the allocation of the limited wireless resources and for respecting the SGCN requirements.

The scope of this work is to design a method for properly allocating the communication resources to the SG nodes having different requirements, by exploiting heterogeneous CTs. In particular, we will focus on latency and data rate requirements. On one hand, we aim at respecting a minimum required data rate and a maximum latency to be assured, but at the same time, we focus on a solution that allows reducing the resource wasting by limiting the allocation of unnecessary resources to the SG nodes. Indeed, it is more preferable that the nodes having lower latency requirements should be supported by CT having intrinsic delays, e.g., satellite communications, and leaving the CTs with lower delays for the nodes requesting a lower latency. However, usually, there is a tradeoff between latency and bandwidth. To this aim, a properly designed cost function is proposed aiming at selecting the priority of each available CT for the different SG nodes.

The proposed method is very effective with respect to other methods because it is simple and it has a low complexity. To the best of our knowledge, the other methods proposed in the literature are not simple resource allocation methods for respecting the SGCN requirements with the given constraints [1–4].

2   The Smart Grid Requirements

There are several types of SG nodes, each one with different uses and requirements. In this section, we will focus on the most important SG node types by describing their requirements.

Advanced Metering Infrastructure (AMI) are a combination of SMs, communications networks, and data management systems, for facilitating and enabling SMs to have two-way communications with the CS [1, 3]. The *Wide Area Situational Awareness* (WASA) nodes monitor the power system across wide geographic areas. Thus, WASA has the important role in SG status and surveillances issues. *Distributed Energy Resources* (DERS) are used for enabling renewable energy resources as a part of the future SG and integrate them into the power grid infrastructure. In addition, DERS work as the power supply resources for emergency usage during outages and disasters are notable. The *Plug in Electrical Vehicle* (PHEV) nodes are beneficial for emissions and fossil fuel energy dependency reductions since they can manage and provide the information about the electrical device charger for electrical vehicles. Finally, the *Distributed Grid Management* (DGM) section allows utilities to remotely monitor and control the parameters in the SG distribution network.

Table 1 summarizes the requirements of the above mentioned SG node types in terms of data rate and latency, where in the first two columns the values as defined by the Utilities Telecom Council (UTC) are reported, while in the other two the values used in this study are reported. UTC has defined such communication requirements based on specific studies for each Smart Grid application, by taking into account also an average number of devices or nodes and the average number of collectors per branch of the network [3].

Table 1. Communication requirements of SG nodes [3]

	Reference data rate [kb/s]	Reference latency [s]	Selected data rate [kb/s]	Selected latency[s]
AMI	500	2–15	500	2
WASA	600–1500	0.02–0.2	1000	0.03
DERS	9.6–56	0.02–15	40	1
PHEV	100	2–300	100	5
DGM	9.6–100	0.1–2	70	0.5

3   The Resource Allocation Cost Function

An evaluation method is needed for properly allocating the communication resources to the different types of nodes in the SG. To this aim, a proper cost function is introduced for managing the resource allocation policy for different nodes with different communication requirements over different communication networks. For achieving these aims, it is needed to define the weights of the most important users Key Performance Indicators, KPIs, and their normalized proportional value in a certain communication network.

Required data rate, or BW, and the delay sensitivity of the SG nodes are two most important KPIs considered in this study. For a certain scenario, the SG nodes having the lowest data rate have the lowest weight and vice versa.

For defining the normalized value, a reference BW and the reference data rate in each different CT are considered. Then, the amount of data rate required to respect the requirements of a certain type of nodes is divided by each CT data rate (for a certain BW in Hz). The same policy is applied to evaluate the weight of the latency based on the delay sensitivity of each type of SG node. Thus, the nodes with lower delay sensitivity have lower weight. The obtained cost function is:

$$CF_{ij} = \frac{\left(W_{bw_{ij}} . N_{bw_{ij}}\right) + \left(W_{delay_{ij}} . N_{delay_{ij}}\right)}{\left(W_{bw_{ij}} + W_{delay_{ij}}\right)} \tag{1}$$

where CF_{ij} is the CF value for the user type i when using the CT j, and $W_{bw_{ij}}$ and $N_{bw_{ij}}$, are the BW weight and normalized value for user type i and CT type j respectively. $W_{delay_{ij}}$ and $N_{delay_{ij}}$ are the delay weight and normalized value for user type i and CT type j, respectively.

It is possible to note that in such a way the communication network with the delay closer to the delay sensitivity of SG nodes has the lower value. Therefore, the node requirements are respected and the resources of the best CT in terms of delay can be allocated to the user with the highest delay sensitivity. The characteristics of the CTs selected for this study are in Table 2.

Table 2. RTT and spectrum efficiency for the three selected communication configuration which are corresponded to the certain CTs [2]

	RTT (ms)	Spectrum efficiency (b/s/Hz)
LTE	10–20 [6]	(1.4 MHz, 64 QAM Modulation)≈3.6 [7]
GSM	150–200 [6]	≈1.36 [8]
(Satellite) LEO	[9, 10] 100–150 <	(8PSK Modulation)≈1.8 [11]

In (1), the BW weight for each node can be defined as $W_{bw_{ij}} = R_{Ni}/M$, where R_{Ni} is the data rate required by the i-th node type, and M equal to $\max\{R_{N1}, R_{N2}, \ldots, R_{Nn}\}$ is maximum requested rate among all the possible node types. The CTs that support a certain type of SG nodes can be ordered by their CF value, where the lowest CF value is the best choice. Therefore, the nodes with the lowest data rate have the lowest weight. The normalized BW value for the node i in the network j is:

$$N_{bw_{ij}} = \frac{R_{Ni}}{PbpsNET_j} \tag{2}$$

where $PbpsNET_j$ is the proportional rate for a certain fixed amount of BW. For example, 1 MHz generates different data rate in different technologies and even in same technology with different modulation scheme (e.g., 5 Mbps in LTE and 1.3 Mbps in GSM). The latency weight for node i can be defined as:

$$W_{delay_{ij}} = 1 - \frac{NWLAT_i}{MAX_{NLAT}} \tag{3}$$

where $NWLAT_i$ is the maximum latency requirement for node i (the last column of the Table 1) and MAX_{NWLAT} is the maximum value among $NWLAT_i$. As mentioned before, the lowest CF value stands for a more efficient allocation. Thus, the node with the higher difference between the delay requirement and delay of the allocated CT, has the higher weight. The normalized latency for node i, when using the network j, can be defined as:

$$N_{delay_{ij}} = 1 - \frac{TotLat_{ij}}{NWLAT_i}. \tag{4}$$

The communication networks having the latency higher than the maximum delay sensitivity of a certain node are ignored since they cannot respect the latency requirements of the node and, based on (4), the normalized value is negative.

Since RTT is the value for the round trip time for each type of communication network and TP is the processing time, considered as 5 ms [12], we will refer in the following to $TotLat_{ij}$ equal to $(RTT + TP)$, as the overall latency value.

As an example, Fig. 1 shows the normalized delay as a function of RTT of CT and the SG node delay sensitivity. As it can be seen, the normalized delay is higher if RTT and SG node delay sensitivity difference is higher and vice versa. This factor helps to define a KPI in which it is preferable to allocate CT with higher RTT to a lower delay sensitivity SG node.

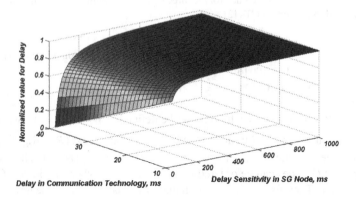

Fig. 1. Normalized value

The Prioritization Method. The prioritization of the SG nodes for respecting their requirements allows selecting the most important nodes. The nodes having a higher priority in the SG will be served earlier. Therefore, it is needed to define the SG goals in terms of KPIs for finding the weight of the SG nodes. Then, it is needed to give a value to each different KPI for a certain node. For a certain type of nodes, giving more importance to a certain KPI depends on how much that node can fulfill that KPI. The intuitive concept we propose is proportional to a quantitative value, as follows: Very high: 5, High: 4, Medium: 3, Low: 2 and Very low: 1.

CF effectiveness in achieving SG goals depends strongly on how SG goals and KPIs are related. The weights given to the KPIs of each node type are used for comparing the behavior of the different types of nodes with respect to certain SG goals. The different types of nodes functionalities, to respect a certain SG goal, are compared among them in order to respect the intuitive and empirical concept in the literature by using the quantitative values. The main goals of SG have been declared in many references, while in the term of KPIs are described in [3].

In the following, a description of different types of services relying on the SGs is done, and a qualitative prioritization is performed. The main goals of SG are: Green Energy, Reliability in power grid, Security in power grid, Outage Avoidance, Users Cooperation, Automated maintenance, Consumption cost minimizing and Disaster Avoidance [3].

Green energy concept in SG is generally defined as energy usage efficiency, decrease using of fossil fuels and try to use the sustainable energy. *Reliability in power grid*, controlling and distribution grid management have the main roles. Increased reliance on renewable improve reliability in associated extreme events. Moreover, its demand side effects in reliability are high [3]. As the technology develops, dependency on the *secure* electricity supplies, transmission and distribution is increased. Grid monitoring and surveillances due to its characteristics has a significant role on SG security. [3].

Outage and blackout avoidance as the result of high consumption or any unpresented faults in the power grid should be considered as an important goal of SG [3, 13]. *User cooperation* is considered as the users' assistance to increase power grid functionality; it is respected on the demand side of the SG. Although, power system status makes information as a feedback to the CS and then CS demand response changes based on it [3]. *Automated maintenance* is an intelligence system whose actions are started automatically at regular intervals to perform maintenance operations. To this aim, SG should monitor all critical components of the power grid [3, 4, 13]. Decreasing the *consumption cost* in the SG platform helps users schedule electrical appliance issues, minimizing variance in power consumption. SG demand side nodes have high effects on it. Although, controlling power status and distributing part have effects on it by detecting the fault over the power grid [3]. *Disaster avoidance* is another important goal of SG achieved through higher rates of survivability following a natural disaster. Beside it, DGM by balancing the power distribution is helpful. Demand side role by communicating with CS on disaster avoidance is notable [3, 4].

Based on the nodes functions in the SG described in the Sect. 2 and the above explained policy to allocate a numerical value to the nodes based on an intuitive understanding of their functions on fulfilling a certain SG goal, the results can be shown in Table 3. As an explanation, the demand side nodes includes AMI, PHEV and even DERS. Large number of users in AMI part, which includes the real users using SMs rather than PHEV, may cause to highlight importance of AMI to fulfill some SG goals rather than PHEV. Although, DERS based on its characteristics has important role to respect to some of SG goals [3, 13].

Table 3. SG nodes weights for different SG goals

Smart Grid Goals	Smart Grid Nodes and Applications				
	AMI	PHEV	WASA	DGM	DERS
Green Energy	4	3	3	4	5
Reliability	3	3	5	5	3
Security	4	4	5	4	3
Outage Avoidance	4	2	5	4	4
Users Cooperation	5	3	3	2	4
Automated Maintenance	1	1	5	3	1
Minimize Consumption Cost	5	4	3	2	3
Disaster Avoidance	3	1	5	4	2
Sum of Weights	29	21	34	28	25
Normalized Weights	0.852941176	0.617647059	1	0.824	0.735

4 Numerical Results

Based on the cost function in (1), the SG nodes priority has been evaluated by a proper simulation framework in MATLAB. The weight of data rate and delay for different SG node are calculated by using (2) and (4).

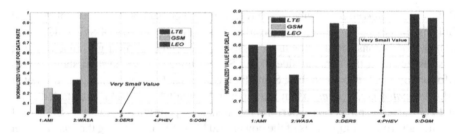

Fig. 2. (a) Data rate normalized value and (b) Delay normalized value

Figure 2a and b show the data rate and delay normalized value respectively for different type of the SG node over three different CTs. As it can be seen, WASA is negative for in case of LEO and GSM since their delay is higher than WASA delay sensitivity. Its negative value by using the CF in (1) can be seen in Fig. 3.

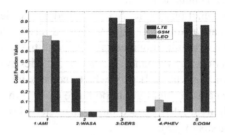

Fig. 3. CF value for different type of the SG node over 3 different communication network

Table 4. Priority table

	First priority	Second priority	Third priority
WASA	LTE	–	–
AMI	LTE	LEO	GSM
DGM	GSM	LEO	LTE
DERS	GSM	LEO	LTE
PHEV	LTE	LEO	GSM

As it is shown in Fig. 3, the nodes like PHEV with lower delay sensitivity have lower delay weight than the other high delay sensitive nodes. The SG nodes in the first column of the Table 4 have been ordered from up to down based on their priority of respecting the SG goals described in the Sect. 3. The priority values in Table 3 and the CF are used jointly for deriving the Table 4. Although LTE is the first priority CT for WASA, AMI and PHEV but at the first step, this resource will be allocated to the SG node with higher priority. This is because a certain CT is not able to support all the nodes, hence, first of all, the highest priority nodes should be supported. Based on the CF values, CFV, and SG node prioritization, SGNP, a priority table is generated. For certain SG node types, the CFV for each different CTs is achieved. The CFV of a certain type of SG node, calculated for all different CTs, are compared and the lowest stands for the most proper technology; based on the CFV and SGNP, the priority table is generated, and used as a criteria to decide which is the best CT for each SG node.

5 Conclusion

Finding a way to allocate the spectrum as the scarce resources to fulfill all smart grid nodes communication requirements in an efficient way is a big challenge. A method was introduced and investigated to properly allocate spectrum of different types of communication technology to a bunch of user types with different characteristics in which all users type meet their communication requirements and avoiding as much as possible the unnecessary allocation of the specific low delay CT resource to a user that is not delay sensitive. Thus a method is introduced based on a proper cost function. Furthermore, the smart grid different nodes types were prioritized based on the smart grid goals and then a defined scenario is investigated based on it.

References

1. Kouhdaragh, V., Tarchi, D., Vanelli-Coralli, A., Corazza, G.E.: Smart meters density effects on the number of collectors in a smart grid. In: EuCNC 2015, pp. 476–481, June–July 2015, Paris, France (2015)
2. Kouhdaragh, V., Tarchi, D., Vanelli-Coralli, A., Corazza, G.E.: Cognitive radio based smart grid networks. In: TIWDC 2013, Genoa, Italy, 23–25 September 2013 (2013)
3. Communications requirements of smart grid technologies, Technical report, U.S. Department of Energy, October 2010
4. Pedersen, A.B., et al.: Facilitating a generic communication interface to distributed energy resources: mapping IEC 61850 to RESTful services. In: SmartGridComm 2010, Gaithersburg, MD, USA, pp. 61–66, October 2010
5. Sauter, M.: From GSM to LTE: An Introduction to Mobile Networks and Mobile Broadband, 1st edn. Wiley, New York (2011)
6. Dahlman, E., Parkvall, S., Skold, J.: 4G: LTE/LTE-Advanced for Mobile Broadband, 2nd edn. Academic Press, New York (2013)
7. Ivanov, A.: TD-LTE and FDD-LTEA basic comparison, Ascom Technical report, January 2012. http://www.ascom.com/en/tems-fdd-lte-vs-td-lte-12.pdf
8. Nadia, A., Aditya, S.K.: Performance analysis of GSM coverage considering spectral efficiency, interference and cell sectoring. Int. J. Eng. Adv. Technol. 2(4), 115–119 (2013)
9. Goyal, R., Kota, S., Jain, R., Fahmy, S., Vandalore, B., Kallaus, J.: Analysis and simulation of delay and buffer requirements of satellite-ATM networks for TCP/IP traffic, OSU Technical report (1998). http://www.cse.wustl.edu/~jain/papers/ftp/satdelay.pdf
10. Kota, S.L., Pahlavan, K., Leppänen, P.A.: Performance analysis of TCP over satellite ATM. In: Kota, S.L., Pahlavan, K., Leppanen, P. (eds.) Broadband Satellite Communications for Internet Access, pp. 355–374. Springer, New York (2004)
11. Piemontese, A., Modenini, A., Colavolpe, G., Alagha, N.: Improving the spectral efficiency of nonlinear satellite systems through time-frequency packing and advanced processing. IEEE Trans. Commun. 61(8), 3404–3412 (2013)

12. Ciet, M., Neve, M., Peeters, E., Quisquater, J.: Parallel FPGA implementation of RSA with residue number systems. In: IEEE MWSCAS 2003, pp. 806–810, Cairo, Egypt (2003)
13. Friedman, N.R.: Distributed energy resources interconnection systems: technology review and research needs, National Renewable Energy Laboratory Technical report, September 2002. http://www.nrel.gov/docs/fy02osti/32459.pdf

Clustering Power Consumption Data in Smart Grid

Kálmán Tornai[✉] and András Oláh

Faculty of Information Technology and Bionics,
Pázmány Péter Catholic University, Budapest, Hungary
tornai.kalman@itk.ppke.hu

Abstract. For power distributors it is very important to have detailed information about the power consumption characteristics of their customers. These information is essential to plan correctly the required amount of energy from power-plants in order to minimize the difference between the demand and supply and to optimize the load of transportation grid as well. For industrial power consumer customers, on the market the actual rate of electric power may depend on their power consumption characteristics. By using intelligent meters and analyzing their behavior, relevant information can be obtained and the consumers can be classified in order to find the best rates for the supplier as well as for the consumer. In this paper, we introduce new results on clustering the consumers. The clustering method is based on forecasting the consumption time series. The numerical results prove that the method is capable of clustering consumers with different consumption patterns with good performance as a result the forecast based method proved to be the a promising tool in real applications.

Keywords: Clustering methods · Consumer clustering · Time series forecast · Feedforward neural network

1 Introduction

Smart power transmission grids can efficiently incorporate the renewable energy resources and also can adaptively manage the balance (supply and demand) of the system. In contrast to present grids, intelligent measurement system and intra-network, two-way communication are integrated into the energy transmission and distribution system.

The usage of smart metering and monitoring devices in the network (both on consumer side and in the transportation network) implies that a huge amount of data is being acquired. These data have to be processed in order to obtain detailed, relevant information about the power consumption characteristics of the customer. Due to the complexity and the amount of data, it is necessary to have sophisticated and innovative algorithms [6, 8]. As a result the new algorithms and the possibilities of the smart grid system imply new applications such as the automatic identification of (i) consumer categories; (ii) outliers in certain groups; (iii) and the misuse of services.

© ICST Institute for Computer Sciences, Social Informatics and Telecommunications Engineering 2017
J. Hu et al. (Eds.): SmartGIFT 2016, LNICST 175, pp. 25–32, 2017.
DOI: 10.1007/978-3-319-47729-9_3

In our previous work we have dealt with the automated classification of consumers, and we have demonstrated that a nonlinear forecast based method is able to determine the correct consumption class of power consumers [10]. However, the categorization of the customers still not fully automated as the number of classes, the initial classification for training has to be determined in advance by technicians.

In this paper, we introduce a new clustering method which is capable of distinguishing different clusters of power consumers with high performance. The method is based on the analysis of error levels of time series forecast. The focus of the method is to differentiate the time series by comparing the underlying processes which are generating them.

The paper will be organized as follows: in Sect. 2, the existing clustering methods and our forecast based classification method will be briefly reviewed. In Sect. 3, the new method will be discussed. In Sect. 4, the performance of the proposed method will be described, finally in Sect. 5, conclusions will be drawn.

2   Related Works

In this section the existing clustering methods and our recently developed classification method are briefly summarized.

2.1   Existing Clustering Methods

Based on [12] the clustering methods can be categorized as follows.

Hierarchical Clustering algorithms organize data into a hierarchical structure. This clustering algorithms can be either agglomerative methods or divisive methods. The agglomerative approach starts with clusters and each of them include exactly one object. If the distance between two clusters is small enough, then the clusters are merged, till the final result is obtained. The divisive clustering proceeds in an opposite way as all the objects belong to one cluster, and the existing clusters are divided into smaller groups [4].

The *squared error based* clustering methods assign a set of objects to clusters without any hierarchical structures. Although the optimal partition can be found by exhaustive search, it is unfeasible because of the computational complexity [7]. Hence heuristic algorithms have been introduced to seek approximate solutions. The k-means algorithm is the most widely used method. It can be used on compact and hyperspherical clusters [13].

Using *mixture Densities-Based Clustering* the data which have to be clustered are assumed to be generated by several probability distributions. So the data can be derived from different density function types (or same density function but different parameters). If the distributions can be determined the problem of clustering is equivalent to estimating the parameters of the underlying models.

Applying the concepts of *graph theory* the nodes of a weighted graph correspond to data points in the pattern space and edges reflect the proximities between each pair of data points. As a result using graph algorithms (such as partitioning, edge cut, etc.) the clustering can be done.

Fuzzy Clustering relaxes the constraint which states that a data object belongs only to one cluster. Therefore an object can belong to all of the clusters with a certain degree of membership. These methods are capable of discovering more sophisticated relations between the clusters and objects [3].

Neural networks-based clustering algorithms mainly consists of self-organizing map (SOM) clusters and adaptive resonance theory (ART). The data is represented by neurons and the strengths of links between the artificial neurons model the connections between the clusters. The final structure of SOM can represent the clusters of the input data. The ART can learn arbitrary input patterns in a stable, fast, and self-organizing way, thus capable of clustering data [1].

2.2 Previous Work – Forecast Based Classification

In this section we briefly summarize the classification scheme which is the basis of the new clustering algorithm. The detailed description and test result can be found in [10,11].

The method exploits the different statistical properties of the power consumption time series acquired from different classes of consumers. The basic idea of the scheme is the following: If an approximator has been trained with time series of a class, then time series of same class can be forecasted with low error rate and time series from any other class can be forecasted with significantly superior error. An individual approximator is trained for every class using representative sequences. After the training phase for the class, the upcoming values of a new sequence will be compared to their forecasted values resulting in a forecast error sequence. Hence the mean of the forecast error will be used as a decision variable to decide the class where the sequence belongs to.

We have considered several linear and nonlinear forecasting methods [9], and the Feedforward Neural Network (FFNN) [2] has been elected as it has the required properties.

3 New Clustering Method Based on Forecast

The novelty of our proposed method is based on the applied approach, where we focus on investigating and recognizing the behavior patterns of the observed consumption data, instead of analyzing the statistical parameters. In our approach the data sequences contain only numerical values and the method is distance based, where the applied metric is the rate of forecast error.

The clustering method is an agglomerative hierarchical clustering method where initially we suppose that each consumer belongs to its own cluster. By applying FFNN our method is model-based as well [2]. The correctness of the empirically built model (which is represented by the artificial neural network) influences the accuracy of the method. However, the model is constructed without any explicit information about the behavior of the process which generates the values of the time series. The artificial neural network is treated as a black box and the training algorithm which extracts the information needed for the model. A training set is constructed containing pairs of input and desired output. During the training phase no assumptions are used regarding the process which generates the time series, only the measured values are supposed to be available. This approach makes our solution different from existing methods.

The data flow of one iteration of the method is depicted by Fig. 1. In each iteration for each existing cluster an approximator is trained using the time series of the cluster. The approximator is used to forecast a forthcoming value of time series using past values of the time series. After training phase finished the approximators are evaluated by all the time series, and the forecasting errors are calculated. Using the forecast error values as distance metric the nearest clusters can be merged, and the iteration is repeated with the reduced number of clusters. This iteration can be repeated until the desired number of clusters are reached, or the cross-detection error rate exceeds a predefined threshold.

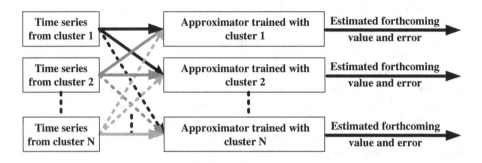

Fig. 1. Algorithmic flow of a single iteration

Formally $\mathbf{x}^{(i)}, (i = 1 \ldots N)$ denotes the power consumption time series for i^{th} customer. The jth cluster of power consumption time series is denoted by $\mathcal{C}^{(j)}, (j = 1 \ldots M)$, the corresponding approximator is denoted by $\Gamma^{(j)}(.)$ respectively. The training set of an approximator is assembled by using samples from the corresponding time series:

$$\tau_{(j)} = \left\{ \left(x_m^{(i)} \ldots x_{m+L}^{(i)}; x_{m+L+1}^{(i)} \right) \right\} \forall i :\in \mathcal{C}^{(j)}, m = 1 \ldots (n^{(i)} - L - 1) \quad (1)$$

where L number of preceding value is being used to estimate the forthcoming value of time series by the approximator. The length of time series i is denoted by $n^{(i)}$.

With the training set the free parameters $(\mathbf{W}^{(j)})$ of the approximators are calculated by learning algorithms. Thus an approximator can be used to forecast the forthcoming values based on L preceding values as follows:

$$\hat{x}_k = \Gamma \left(\mathbf{W}, x_{k-1,\ldots,k-L} \right). \quad (2)$$

When the $\hat{x}_k^{(i)}$ estimated value is determined, it can be compared with the real $x_k^{(i)}$ value, then the forecast error can be calculated. The forecast error for a single time series i with approximator j is

$$\sqrt{\frac{1}{n^{(i)}} \sum_{k=1}^{n^{(i)}} \left(x_k^{(i)} - \hat{x}_k^{(i)}\right)^2}, k = (L+1)\ldots n^{(i)} \tag{3}$$

For each i time series and j cluster the previous value can be determined as a result an $\mathbf{E} \in \mathbb{R}^{M \times N}$ matrix of error values can be constructed. The values of $\mathbf{E}$ can be considered as the distance between time series of different clusters. The next step of the method is to find two candidate clusters to be merged. In order to do so, the minimal forecast error value is sought. If a time series can be forecasted with two or more approximators with significantly lower error rate than the mean of forecast errors of other approximators, it is presumed that the clusters can be merged. After merge is performed a new iteration started with the reduced number of clusters. When minimal error rate exceeds any of intra-cluster forecast error rate then the iteration is terminated as the final result is available.

4   Performance Analysis

In this section the data model, test environment and the performance of the proposed method is introduced.

4.1   Data Models

The clustering method were tested by artificially generated time series data and real power consumption time series as well. Two approaches were used to generate time series: (i) autoregressive (AR) processes; (ii) a realistic model, where the consumer's consumption data were constructed as the sum of statistically independent consumption data of electric appliances.

AR Model for Consumption Time Series. Exemplary parameters for AR(5) processes (for different classes) are the rows of matrix (4). Using these values such time series can be generated, which have different statistical properties.

$$\begin{pmatrix} 0.7 & -0.2 & 0.1 & -0.2 & 0.1 \\ 0.3 & 0.7 & 0.3 & -0.4 & -0.2 \\ 0.4 & 0.2 & 0.1 & 0.2 & -0.1 \end{pmatrix} \tag{4}$$

Markovian Model for Consumption Time Series. We have generated close-to-real consumption time series using the following model. The consumption of individual appliances is assumed to be a two-state Markovian process which can model the time dependencies of the time series. Formally, a single appliance is modeled with a discrete random variable

$$X^{\text{Markovian}} \in \{0, h\}, \tag{5}$$

where 0 means that the consumer is switched off, and h is the energy level of the powered device. The parameters of the Markovian model were fitted on real measurements coming from the REDD database [5]. We have investigated the correctness of the model comparing the autocorrelation values of generated data and real power consumption time series. The generated values has similar long-term correlations, however the artificial data do not have the characteristics of daily periodicity.[1]

The total power consumption of a customer is modeled by the sum of numerous, individual Markovian processes. Each class contains different types of total power consumption data which are constructed from different types of appliances.

4.2 Real Consumption Data

These data were obtained from a large Central-European electricity distribution company. The consumption were measured at 150 sites for one year in 2009. Each value in the time series represent aggregated power consumption for one hour. As the actual data is trade secret it has been normalized by the company and personal information was removed as well.

4.3 Performance Results

The simulations has been carried out in Matlab environment. We have repeated all test for several times to have averaged result - in order to eliminate the preferential cases. For each test data the correct clusters are known prior to the test due to the method of generating the data. Hence the performance of the proposed method can be evaluated by comparing the results of the method with the known information. In the tests 30 time series were used, which are from three different classes.

Figure 2 demonstrate the forecast error values of the initial and termination states of the clustering method on Markovian model generated artificial time series. At the initial state each customer belongs to its own cluster, and the inter-cluster forecast errors are investigated. As all the neighboring clusters are merged the final result is available. It is observable that the lowest error rates are always at the approximator of class where the time series belongs to.

The rates of correct assignments of times series are summarized on Fig. 3. The results are averaged and tested with different amount of time series and type of data model. As figure illustrates the forecast based clustering method is capable of correct clustering both AR and On/Off model generated artificial

---

[1] Paper with more details on results is submitted for publication.

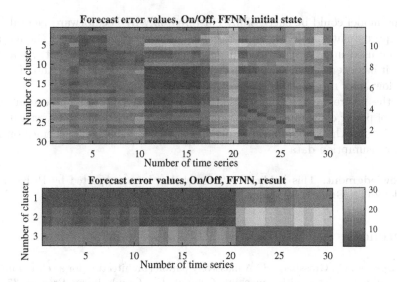

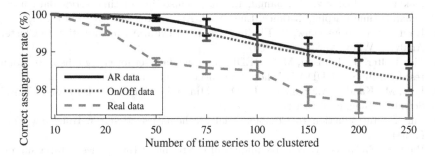

Fig. 2. (a) Forecast error values after the (a) first iteration, before any if cluster merging is done; (b) clustering algorithm is terminated

Fig. 3. Averaged correct assignment rates for clustering method on different amount of time series using different data model

time series as well as real power consumption times series at least with 97 % accuracy. Further investigation required regarding to impact of the number of time series, and the number of expected clusters on the performance.

5   Conclusion

In this paper nonlinear approximators have been applied in order to cluster different electricity consumers that have the same first and second order statistics but have different distributions and time dependencies. Our scheme has been tested by different consumption models, from which the more realistic bottom-up

Markov model could be successfully clustered only by Feedforward Neural Network approximator. Furthermore, it has been investigated whether the method could be used in case of real-world power consumption data or not. As a final result, it has been shown that feedforward neural network based method is capable of low error rate clustering in real applications using real data.

In the future we are going to extend the validation tests on several other classes of power consumption data. Furthermore, we are going to investigate the capabilities of the combination of classification and clustering schemes on real power consumption data.

Acknowledgment. This publication/research has been supported by PPKE KAP 15-084-1.2-ITK Grant. This source of support is gratefully acknowledged.

References

1. Carpenter, G., Grossberg, S.: A massively parallel architecture for a self-organizing neural pattern recognition machine. Comput. Vis. Graph. Image Process. **37**, 54–115 (1987)
2. Haykin, S.: Neural Networks, a Comprehensive Foundation, 3rd edn. Prentince Hall, Pearson (2008)
3. Izakian, H., Pedrycz, W., Jamal, I.: Fuzzy clustering of time series data using dynamic time warping distance. Eng. Appl. Artif. Intell. **39**, 235–244 (2015)
4. Kaufman, L., Rousseeuw, P.: Finding Groups in Data: An Introduction to Cluster Analysis. Wiley, New York (1990)
5. Zico Kolter, J., Johnson, M.J.: REDD: a public data set for energy disaggregation research. In: SustKDD Workshop, San Diego, California (2011)
6. Last, M., Kandel, A., Bunke, H.L.: Data Mining in Time Series Databases. World Scientific, Singapore (2004)
7. Liu, G.: Introduction to Combinatorial Mathematics. Mc-Graw-Hill, New York (1968)
8. Rani, S., Sikka, G.: Recent techniques of clustering of time series data: a survey. Int. J. Comput. Appl. **52**(15), 1–9 (2012)
9. Sruthi, J., Helen Catherine, R.L.: A review on electrical load forecasting in energy management. IJISET - Int. J. Innov. Sci Eng. Technol. **2**(3), 670–676 (2015)
10. Tornai, K., et al.: Novel consumer classification scheme for smart grids. In: 2014 European Conference on Smart Objects, Systems and Technologies (Smart SysTech), July 2014
11. Tornai, K., Oláh, A., Lőrincz, M.: Forecast based classification for power consumption data. In: International Conference on Intelligent Green Building & Smart Grid (2016)
12. Xu, D., Tian, Y.: A comprehensive survey of clustering algorithms. Ann. Data Sci. **2**(2), 165–193 (2015)
13. Zhang, B., Srihari, S.N.: Fast k-nearest neighbor classification using cluster-based trees. IEEE Trans. Pattern Anal. Mach. Intell. **26**(4), 525–528 (2004)

Using a Cost Function to Choose the Best Communication Technology for Fulfilling the Smart Meters Communication Requirements

Vahid Kouhdaragh[✉], Alessandro Vanelli-Coralli, and Daniele Tarchi

Department of Electrical, Electronic and Information Engineering,
University of Bologna, Bologna, Italy
{vahid.kouhdaragh2,alessandro.vanelli,daniele.tarchi}@unibo.it

Abstract. The conventional power grids are not efficient today so their ineffective functions need to be managed in a more effective way. This goal is at the base of the Smart Grid (SG) concept, to present intelligence in the energy grid. Among several SG nodes, smart meters (SMs) work in the demand side of the power grids and their number is much increasing over time. A SM is a SG device which records electric energy consumption in certain time intervals for communicating the information to the SG Control Station (CS) through the collectors and aggregators. The wireless communications have a significant role in SG functions. Hence, the spectrum scarcity due to the growing number of users is becoming a significant problem. Thus introducing an algorithm to avoid facing with spectrum scarcity by defining a Cost Function (CF) is helpful in sense of two issues. First, all SMs meet their communication requirements in the sense of delay sensitivity and data rate. The second goal is to avoid as much as possible the unnecessary allocation of the specific low delay Communication Technologies (CTs) spectrum to the SMs that are not delay sensitive. It is more preferable to support the less delay sensitive users with satellite or the CTs (with certain communication configurations such as high latency specification) and keep the low delay communication spectrum such as LTE for the users that are more delay sensitive. This paper is focused to introduce a method to achieve these goals.

Keywords: Smart meters · Delay sensitivity · Communication configuration · Cost function · Spectral efficiency · High delay networks

1 Introduction

The conventional power grids are no more effective and new models are needed to accomplish current needs effectively: Smart Grids (SG). There are different types of SG devices and nodes, whose number is increasing continuously. They report electrical power information details to Control Stations (CS) through the collectors. Then demands responses such as dynamic consumption cost and controlling commands are prepared by CS to be sent to the interested nodes. Smart Meters and its network structure, Advanced Metering Infrastructure (AMI) are considered as the backbone of the SG [1–3]. Network requirements for meter reading applications vary

© ICST Institute for Computer Sciences, Social Informatics and Telecommunications Engineering 2017
J. Hu et al. (Eds.): SmartGIFT 2016, LNICST 175, pp. 33–42, 2017.
DOI: 10.1007/978-3-319-47729-9_4

depending on service types but, defining an assigned period time and message size to each SMs make the communication network more efficient [1, 4]. Scheduled meter interval reading provides the capability to collect usage information and transfer it from a meter to a collector in AMI system several times a day. Accumulated SMs data in collectors is sent to the CS by using different CTs [1]. However, spectrum scarcity for future machine to machine (M2 M) communication is becoming a big challenge [1, 3, 6]. Several of these M2 M nodes are very delay sensitive. Besides, it is needed to find a proper way to allocate this limited source to fulfill the SMs communication requirements which are not delay sensitive but the huge number of hem will be used in future SG infrastructure. Thus, the unnecessary allocation of the specific CT (with a certain Communication Configuration(s); **C.Conf(s)**) spectrum with low Round Trip Time, RTT, or high delay communication network allocation to the users that are not delay sensitive is an inefficient allocation. On the other hand, the mentioned spectrum can be conserved for a more delay sensitive user. Beside it, the same amount of bandwidth, BW, is proportional to different data rate in different configuration of CTs. Low delay sensitivity and needed bit rate for the SMs data accumulated in a collector makes it difficult to design an efficient communication network for it. Lots of studies have been done on the SG communication network, but still a lack of resource allocating method based on the SMs nodes communication requirements and communication network characteristics has been remained as an open issue [1, 3, 5].

The scope of this work is investigating a method to properly allocate resources of the different CTs with different C.Confs to the SMs having two aims. First, all users type come across their communication requirements in the sense of delay sensitivity and data rate. The second aim is to avoid as much as probable the unnecessary allocation of the specific low delay CT (with a certain C.Conf) spectrum to a user that is not delay sensitive. Thus a proper method for comparing some different CTs (with different C.Conf) functionality that supports a certain number of SMs should be defined. Defining a Cost Function (CF) which its value is an indication of the desirability of communication network to support a certain number of SMs, with respect to the SMs communication requirements and characteristics of the different CTs seems to be a good approach to deal with these issues. It is a way to quantify the desirability of each C.Conf corresponding to a certain CTs and obtain numeric criteria to compare the effectiveness of CTs in fulfilling the SMs requirements, with each other (to make a most efficient matched between nodes communication requirements and C.Config C.o.to a certain CT). The CF value for each node type and C.Config C.o.to a certain CT can also clarify a way to find the minimum number of the certain nodes that can be supported by a certain C.Config C.o.to a certain CT among the different CTs. Thus the total system can be benefitted from the Heterogeneous network advantages. This approach is not discussed in this paper and can be considered as another application of this method. However this can never be achieved totally but, the Cost Function, CF, designation is an effort to meet the above goals as much as possible. The two most important communication requirements (CR) for SMs are data rate and delay [5, 6] which are considered as the Key Performance Indicators (KPIs) for the mentioned CF. It should be mentioned that the two different KPIs, Reliability and Security are not included in the CF and are not discussed in this paper due to importance of delay and data rate as the main KPIs . The way to define weights and formulized the normalized value for the CF are introduced in this paper. Then by considering different CTs (C.Confs) and the CF value, the better CT based on the introduced goals is chosen.

2 Smart Meters Communication Requirements

SMs usually generate 1000 Bits in each 5–20 min though it differs for SMs and depends on the nature of the buildings they are used [1, 6]. In some factories with plenty of electrical devices each SMs reports the data every 5 min [1]. For some houses with lower electrical devices the report generation interval is reduced to 20 min; thus, 15 min reporting period for each SMs, appears to be an acceptable and truthful reporting time for the urban and suburban users [1, 6]. The more details can be seen in Table 1.

Table 1. Smart Meters communication requirements [1, 6]

Node	Reporting time period	Data size (*Dps*)	Delay sensitivity (DL_s) in Second
SM (in AMI)	Every 15 min	1000 bits	2-15 [6] (DL_s = 15 s, selected value)

Assume there are N_{sm} SMs that generate a definite amount of data, *Dpsbits*, every 15 min. It is assumed that the size of the packets generated by each SM is constant and equal to 1000 bits [1, 6]. For 4000 SMs in AMI, a collector receives data from 200 numbers of SMs in average in a second. Thus a collector receives the data every 0.005 s from one SM in average. It is shown in [1, 7] 500 kbps is quite good enough to support 4000 SMs [1, 7]. Referring to literature, for 4000 SMs, 500 Kbps data rate of a single aggregator can supports these amounts of SMs. Moreover, it is shown for a certain number of SMs, the buffered size data, Buf_{SM}, in a collector can be calculated by using (1) [1].

$$Buf_{SM} = s \times N_{sm} \tag{1}$$

where Buf_{SM}, s and N_{sm} are, respectively, the buffered data in aggregator in 1 s (*Kbits*), the line slip and the number of SMs. Based on the literature, s, can be easily calculated as following:

$$s = \frac{500\,\textbf{Kbps}}{4000} = 0.125\,\textbf{Kbps}, \textbf{thus}: \textbf{Buf}_{SM} = 0.125 \times N_{sm}\,\textbf{Kbps}[1]$$

Therefore, by having the number of SMs, the approximately accumulated data which an aggregator collects for an AMI system can be achieved. These aggregators can be supported by a certain type of CTs with a certain configuration. The aggregated data should be transferred to the Control Station (CS) through a communication link [1, 4]. Due to the delay sensitivity of aggregated data, and different CTs configurations and characteristics, it is important that an appropriate C.Conf is chosen. It is discussed in the next section.

3 System Model

3.1 Communication Network Infrastructure and Communication Technologies Configurations

AMI is considered as the backbone of the SG, dealing with smart meters and its network structure. Their important role on the SG demand side is notable. AMI can be seen as a combination of SMs, communications networks, and data management systems, for facilitating and enabling SMs to have two-way communications with the CS [1, 3, 4]. The SMs data are collected by aggregators or collectors and then it is sent to the Base Station (BS) and CS. Two different communication network infrastructures are used in this paper. At first one which is shown in Fig. 1, the SMs data are collected by collectors and then the collectors transfer the data to the CS through a wireless terrestrial communication network such as LTE or GSM or WiMAX [1, 4, 7].

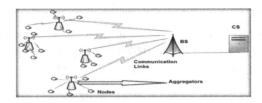

Fig. 1. Communication network structure for AMI using terrestrial wireless network

Second communication infrastructure which is used in this paper is shown in Fig. 2. In this infrastructure, SMs data are collected by collectors and then they transfer the data through the satellite communication link to the BS and CS. Respect to the satellite's orbit distance to the earth the propagation delay and Round Trip Time is increased.

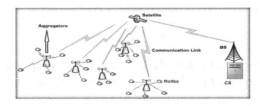

Fig. 2. Communication network structure for AMI using satellite communication network

The collectors in all of these types of network infrastructure use different type of CT with a certain C.Confs that is shown in Table 3. The abbreviations denotation which are uses in Table 3 are shown in Table 2. By considering BW = 1 MHz, the approximated data rate for each C.Conf is achieved by using the Eq. (2) and the spectral efficiency (SE) of configuration j (Sef_j) corresponding to a CT. The Eq. (2) is used in the CF part and defining the normalized value for both data rate and delay for a certain type of C.Conf. These CTs with the defined configuration are used to transfer collector's data to the CS.

$$DRconfig_j = BW \times Sef_j \qquad (2)$$

$j \in \{1,2,3,\ldots,F\}$ in which F is the maximum number of communications configurations which corresponds to a certain CTs. Respect to Table 3 for BW = 1 MHz, $DRconfig_j = \{DRconfig_1, \ldots, DRconfig_5\}$ are Co.to LTE, GSM, Satellite(LEO, MEO and GEO) respectively.

Table 2. Abbreviation for Table 3

PD (Pdl_j)	PT($Pros_j$)	Co.to	M.S
Propagation Delay	**Processing Time**	Configuration is Corresponded to	Modulation Scheme

Table 3. Different Configurations (Config. j) corresponding to the CTs [8–11]

	Config. 1	Config. 2	Config. 3	Config. 4	Config. 5
SE bits/Hz	4	1.35	1.8	1.2	1.07
M.S	64 QAM	GMSK	8PSK	DVB-S2, Extensions APSK	4PSK OR PSK
PD, msec	5	5	25	150	400
PT, msec	5	5	5	5	5
Co.to	LTE	GSM	LEO	MEO	GEO

3.2 Cost Function and Key Performance Indicators Definition

To reach the described aims, it is needed to define the weights of the most important users Key Performance Indicators, KPIs, and their normalized proportional value in a certain communication network. It depends on the CF definition, which lower or higher CF value is favorable. In this paper, for the communication CF part, the weight and normalized value are defined in a way which the lower value is more desirable. In addition, it should be mentioned that for all the weights and normalized values; WDR_{sm}, WDE_{sm}, NDR_j and NDE_j which are data rate weight, delay weight, data rate normalized value and delay normalized value respectively, it should be always: $0 < WDR_{sm}, WDE_{sm}, NDR_j, NDE_j \leq 1$. The weights can be chosen by designer to highlight the effect of a certain KPI. Required data rate and the delay sensitivity of the SMs are two most important KPIs which are considered in this study. It is because of this fact that the user who has low data rate requirements would be fulfilled with lower data rate (or even lower bandwidth) which is desirable. For defining the normalized value, it is considered a certain value of BW (Hz) and its proportional data rate in each different CT with a certain C.Confs. Then the amount of data rate required to fulfill the SMs communication requirements is divided by data rate of each C.Conf corresponded to a CT. The normalized value for data rate calculation formula is given in Eq. (3) in which $Buf_{SM}(1)$ is the data size is buffered in the aggregators every one second.

$$NDR_{sj} = \frac{Buf_{SM}}{DRconfig_j} \qquad (3)$$

By using (4), the total delay to send a bulk of buffered data to the CS is calculated in which Pdl_j, $Pros_j$ and α are propagation delay; if $\alpha = 1$ is assigned; (C.Conf j Co.to a CT), processing delay and propagation delay effect coefficient respectively. In (4), α is the coefficient can be assigned by designer to highlight the propagation delay effect, especially when the satellite communication is considered as the communication network. To define the normalized delay values, (5) is used in which the communication network with inherent delay closer to the delay sensitivity of SMs, have lower value. Therefore, the node requirements are fulfilled and the spectrum of a low delay CT can be allocated to the user with high delay sensitivity. It should be mentioned the sum of $(\alpha.Pdl_j)$, accumulated data transferring time and processing time cannot be more than SMs delay sensitivity (as it is obvious from (5), the nominator should be more than the denominator, because as it is mentioned all the weights and normalized value are higher than 0). Thus by considering the highest propagation delay corresponding to GEO, maximum $\alpha \approx 30$ is considered (30 times more than PD using GEO satellite and data transferring time will be approximately near 15 s that is the SM delay sensitivity, Table 1). Therefore, $1 < \alpha < 30$ and in the mentioned design, it is considered $\alpha = 10$.

$$TotLat_{sj} = \left(\frac{Buf_{SM}}{DRconfig_j} \right) + \left(\alpha.Pdl_j + Pros_j \right), \qquad (4)$$

$$NDE_{sj} = 1 - \frac{TotLat_{sj}}{DL_s} \qquad (5)$$

Figure 3 shows the delay normalized value for SMs when $\alpha = 1$ and $\alpha = 10$. As it can be seen, the normalize values are decreasing by increasing the network delay in both cases. Based on the policy to define the delay normalized values it can be seen that the networks which their latency is closer to the SMs delay sensitivity, their delay normalized value is lower that results to lower the CF value and thus the low delay sensitive

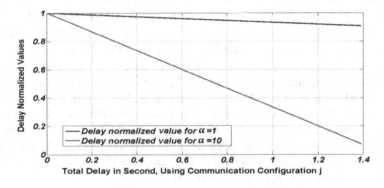

Fig. 3. Normalized value for delay vs different amount of delay for SM, Eq. (7)

node such as SMs has lower CF value when is supported by a high delay communication network which is favorable based on the introduced policy in this paper. Moreover, by considering $\alpha = 10$, the effect of the propagation delay is more significant and the lower value for the normalized value than situation that $\alpha = 1$ for the same network delay is achieved which results in having lower CF in which the network delay effect has been highlighted more.

Because of importance of using high latency CTs with a certain C.Conf for supporting low delay sensitive nodes like SMs, and also considering the effect of data rate of a certain configuration of the CT in the defined formula for delay normalized value, the WDR_{sm} and WDE_{sm} are assumed 0.01 and 1 respectively (The effect of the data rate weight should be decreased significantly because it is considered in computing the normalized value for delay; (4) and (5)). By defining the weights and normalized value and using the CF (6) the values for different number of SMs which is supported by different C.Conf j (corresponded to CTs) are achieved and the best CTs choice for a certain number of SMs can be achieved.

$$CF_j = \frac{\left(WDR_{sm}.NDR_{sj}\right) + \left(WDE_{sm}.NDE_{sj}\right)}{\left(WDR_{sm} + WDE_{sm}\right)} \quad j \in \{1, 2, 3, \ldots, F\} \tag{6}$$

Using Eqs. (3)–(6), the finalize formula for the CF is obtained as Eq. (7).

$$CF_j = \frac{\left(\dfrac{WDR_{sm}.Buf_{SM}}{DRconfig_j} + WDE_{sm}.\left(1 - \dfrac{\left(\dfrac{Buf_{SM}}{DRconfig_j}\right) + \left(\alpha.Pdl_j + Pros_j\right)}{DL_s}\right)\right)}{\left(WDR_{sm} + WDE_{sm}\right)} \tag{7}$$

Figure 4 show in details how the CF works. Definition of SMs communication requirements in terms of delay sensitivity and data rate is the first step. Then respects to the CTs configuration details, the normalized value are defined. Also the weights of the KPIs are defined. Then by using the CF formula, the CF values for a certain number of SMs which are supported by different CT j are determined. Then the CTs are put in the order in priority table with respect to the CF values which are obtained for a defined number of SMs. Functionality of the proposed CF is discussed in the next part. Reliability is another KPIs that can be measured for a certain C.Config C.o.to a certain CT supporting a certain node by finding Mismatch Probability (MMPR) and it depends to the node delay sensitivity and generating rate of the node and may a high MMPR would be desirable for a certain node type and not be desirable for the different node type [6, 12]. The other KPI is security that can be measured by using a numerical method to determine the security desirability of each C.Config C.o.to a certain CT. But as it was mentioned just two most important KPIs, Delay and Data rate are included in the CF in this paper.

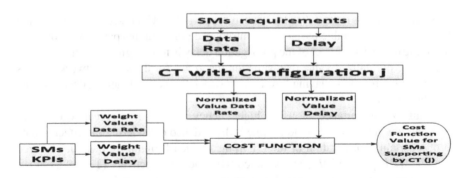

Fig. 4. Cost Function flowcharts

4   Numerical Results and Discussions

Based on the proposed CF and KPIs weights and normalized value the CF values for
maximum 10000 SMs are achieved for different C.Conf which each is corresponded to
a certain CT. The results are shown in Fig. 5.

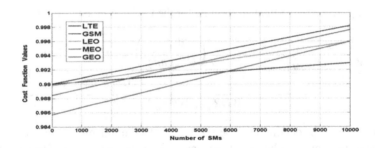

Fig. 5.   CF for the communication configurations that are corresponded to certain CT vs #SMs

For an AMI infrastructure with 4000 SMs, as it is seen in Fig. 5 the C.Conf which
is Co.to satellite communication using GEO satellites is the best choice due to its lowest
value for 4000 SMs. The Table 4 shows the CTs are put in the order in priority table
with respect to the CF values which are obtained for 4000 SMs.

Table 4.   Priority table for 4000 SMs (for a single AMI)

	Priority 1	Priority 2	Priority 3	Priority 4	Priority 5
Co.to	GEO	LTE	MEO	LEO	GSM

When the SMs numbers increases, the data rate normalized value increase as well.
Moreover, the delay normalize value decrease because the more buffered data needs
more time to be transferred to the CS (refer to Eqs. (4 and 5)). The LTE can be the best
choice for the $N_{sm} > 6000$ because higher buffered data by SMs have a significant effect
in both data rate and delay normalized value. As it is shown in Fig. 5, GSM has the

highest CF value among the others and does not have any advantages over the others based on the proposed methods. The configuration 3 (Co.to LEO) due to its higher SE, shows the better functionality than the configuration 5 (Co.to GEO) for $N_{sm} > 10000$. The configuration 4 (Co.to MEO) shows the better functionality than configuration 3 (Co.to LEO) due to its higher propagation delay but, for $N_{sm} > 5000$ their functionality got inversed due to higher SE in configuration 3 (Co.to LEO) than configuration 4 (Co.to MEO). For 8000 SMs (each AMI, 4000 SM) the priorities from 1th to 5th are Co.to LTE, GEO, LEO, MEO and GSM respectively.

5   Conclusion

Finding a way to allocate the spectrum as the scarce resources to fulfill all Smart Meters communication requirements in an effective way is a big challenge. A technique was introduced and investigated to properly choose the best CTs corresponding to a certain communication configurations with different characteristics in which all SMs meet their communication requirements and avoiding as much as probable the unnecessary allocation of the specific low delay CT with a certain communications configurations to the SMs which is not delay sensitive node. Thus a method is introduced based on a proper cost function. The numerical results were achieved based on the proposed CF and it was implemented for maximum 10000 SMs over 5 different communication configurations and the results were analyzed.

References

1. Kouhdaragh, V., Tarchi, D., Vanelli-Coralli, A., Corazza, G.E.: Smart meters density effects on the number of collectors in a Smart Grid. In: 2015 European Conference on Networks and Communications (EuCNC), pp. 476–481, 29 June 2015–2 July 2015
2. Kuzlu, M., Pipattanasomporn, M., Rahman, S.: Communication network requirements for major smart grid applications in HAN. NAN WAN. Comput. Netw. **67**, 74–88 (2014)
3. DNV KEMA Energy and Sustainability: The Critical Need for Smart Meter Standards, a global perspective. Arnhem, The Netherlands (2012)
4. Xiao, Y.: Communication and network in Smart Grid. Department of Computer Science, University of Alabama, 25 April 2012. by CRC Press Reference - 325 Pages - 115 B/W Illustrations, ISBN 9781439878736 - April 25, 2012
5. Mäder, A., Rost, P., Staehle, D.: The Challenge of M2 M Communications for the Cellular Radio Access Network. EuroView 2011, Wurzburg, Germany (2011)
6. U.S. Department of Energy: Communications Requirements of Smart Grid Technologies. Report, 5 October 2010
7. Rengaraju, P., Lung, C.-H., Srinivasan, A.: Communication requirements and analysis of distribution networks using WiMAX technology for smart grids. In: 8th International Wireless Communications and Mobile Computing Conference (IWCMC), pp. 666–670, 27–31 August 2012
8. Sauter, M.: From GSM to LTE", 1th ed. West Sussex, UK, 2011, Chap. 9, 9.1–9.20 (2011)
9. Minoli, D.: Innovations in satellite communication and satellite technology. In: 1th Edn., ISBN: 978-1-118-98405-5, 448 pages, April 2015

10. Piemontese, A., Modenini, A., Colavolpe, G., Alagha, N.: Improving the spectral efficiency of nonlinear satellite systems through time-frequency packing and advanced processing. IEEE Trans. Commun. **61**, 3404–3412 (2013)
11. Raja Rao, K.N.: Fundamentals Of Satellite Communication, January 2004. ISBN 10:8120324013
12. Shawky, A., Olsen, R., Pedersen, J., Schwefel, H.P.: Class-based context quality optimization for context management frameworks. In: 2012 21st International Conference on Computer Communications and Networks (ICCCN), pp. 1–5, Munich 2012

Assessing Loss Event Frequencies of Smart Grid Cyber Threats: Encoding Flexibility into FAIR Using Bayesian Network Approach

Anhtuan Le[1(✉)], Yue Chen[1], Kok Keong Chai[1], Alexandr Vasenev[2], and Lorena Montoya[2]

[1] School of Electric Engineering, Queen Mary University of London,
Mile End, London E1 4NS, UK
{a.le,yue.chen,michael.chai}@qmul.ac.uk
[2] University of Twente, Drienerlolaan 5, 7522 NB Enschede, The Netherlands
{a.vaseneva,l.montoya}@utwente.nl

Abstract. Assessing loss event frequencies (LEF) of smart grid cyber threats is essential for planning cost-effective countermeasures. Factor Analysis of Information Risk (FAIR) is a well-known framework that can be applied to consider threats in a structured manner by using look-up tables related to a taxonomy of threat parameters. This paper proposes a method for constructing a Bayesian network that extends FAIR, for obtaining quantitative LEF results of high granularity, by means of a traceable and repeatable process, even for fuzzy input. Moreover, the proposed encoding enables sensitivity analysis to show how changes in fuzzy input contribute to the LEF. Finally, the method can highlight the most influential elements of a particular threat to help plan countermeasures better. The numerical results of applying the method to a smart grid show that our Bayesian model can not only provide evaluation consistent with FAIR, but also supports more flexible input, more granular output, as well as illustrates how individual threat components contribute to the LEF.

Keywords: Cyber threat · Loss event frequency · Threat assessment

1 Introduction

Researchers have recently addressed the challenges of protecting the smart grid from cyber security threats, which can significantly impact the power system as well as human life [1]. Several authors have pointed out the threats to smart grids, especially from the viewpoint of information security [1–4]. The need to assess such threats for the planning of security resources and mitigation plans is being increasingly recognized. Existing approaches include quantitative, qualitative, and hybrid assessments. The ultimate goal of the quantitative approach is to utilize probability theory and statistics to assign numerical probability values to threat likelihood [5]. While these methods can provide clear guidance about the threats, they are very difficult to implement and evaluate [6]. On the other hand, the qualitative techniques rely on a systematic expert analysis for providing qualitative output rather than a quantitative one [7]. Their main advantage is that they involve reliable expert reasoning, however, in

© ICST Institute for Computer Sciences, Social Informatics and Telecommunications Engineering 2017
J. Hu et al. (Eds.): SmartGIFT 2016, LNICST 175, pp. 43–51, 2017.
DOI: 10.1007/978-3-319-47729-9_5

many cases, the output is not detailed enough to help take clear decisions [8]. Recently, several hybrid models were proposed that combine the quantitative and qualitative methods and eliminate their weaknesses. Among the hybrid approaches, the Factor Analysis of Information Risk (FAIR) framework [9] is well-known and applicable to many risk and threat assessment situations, due to its effective yet simple practical guidelines. Instead of purely qualitative analysis, FAIR assesses threats by means of the Loss Event Frequency (LEF) concept using a five-point scale. However, if many threats need to be assessed, it is necessary to extend this scale further to differentiate threats in the same group. This is particularly relevant for smart grids as these are highly complex systems that are exposed to a wide variety of threats from diverse threat actors.

In this paper, we propose a method to construct a Bayesian network model based on FAIRs LEF concept and a look-up table for supporting the analysis in the context of the IRENE project (i.e. the resilience of the energy grid) [5]. The proposed Bayesian model is consistent with the FAIR's look-up tables. However, the difference between the two is that our model provides a numerical output instead of a categorical one. Moreover, the method has several advantages due to its design as it:

– *Supports ranking threats in the same group.* By providing systems managers with a numerical output, a threat's LEF can be contrasted against other threats in the same group. The managers can therefore make better decisions regarding security countermeasures and mitigation plans;
– *Generates an output even from fuzzy inputs*, for instance, when experts do not fully agree on specific threat parameters;
– *Illustrates how changes* in the input data propagate through the network and contribute to the output;
– *Points out the most influential factor* that, if lowered, could decrease the overall LEF by a greater margin than the others.

The remainder of this paper is organized as follows. Section 2 introduces our model to transform the FAIR framework to the Bayesian network reasoning. Section 3, presents experimental results and then discusses how to consider several threats in a smart grid configuration. Finally, Sect. 4 presents the conclusions.

2   Proposed Model

2.1   The FAIR Framework

This paper addresses how threats can be assessed using FAIR's LEF concept using a reasoning structure for a number of threat factors. These factors include (i) Contact (C): the frequency within a defined timeframe that the attackers will come in contact with the asset, (ii) Action (A): the probability that an attacker will act against an asset once contact occurs, (iii) Threat Capability (Tcap): the probable level of force that an attacker is capable of applying against an asset, and (iv) Control (i.e. Resistance) Strength (CS): the strength of a control compared to a baseline measure of force. The reasoning structure between these factors is presented in Fig. 1. These FAIR constructs can be projected to the risk assessment constructs from NIST 800-30 Guide for Conducting Risk Assessments as shown in [10].

FAIR encodes each threat factor by means of a five-point scale (i.e. Very Low, Low, Moderate, High, and Very High). It also provides a reference for estimating the state of the input factors. A more detailed explanation on how to derive FAIR's input states can be found in [11]. After the cause states (Fig. 1, left) are established, FAIR provides reasoning tables to look up the effect state. The look-up table in Fig. 1 shows how the "LEF" factor can be derived from the "Threat Event Frequency" (TEF) and the "Vulnerability" (V) factors. If values for Contact, Action, Control Strength, and Threat Capability are provided, the TEF and Vulnerability states can be derived, leading to the LEF state.

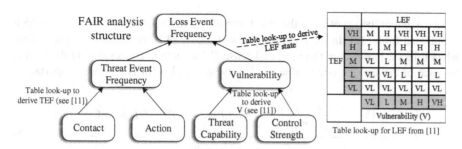

Fig. 1. The FAIR model's LEF analysis structure (left) and the look-up table for deriving the LEF state (right)

2.2 Bayesian Network Approach to Transform a Structural Analysis

This paper applied the method proposed in [12], and developed a method to construct the Bayesian Conditional Probability Table (CPT) of an effect based on the fuzzy relations of the causes that lead to it. The transformed model is given a Bayesian reasoning structure with n **causes** that lead to an **effect**. The **causes** and the **effect** all can have m states, represented as state $1, 2, ..., m$. For example, in the FAIR analysis, the TEF can be considered as the effect, while the Contact and Action are the causes. Each of these factors have 5 states, which are [VL, L, M, H, VH]; VH is the highest level state (level 5), while VL is the lowest level state (level 1). One assumes that each cause i affects the effect through the *individual effect vector* $[r_{i1}\ r_{i2} \dots r_{im}]$, meaning that, if the state of cause i is j, then it will contribute r_{ij} percent to the event that the effect has the highest state (state m) on. On the other hand, one assumes that the relationships between the causes and the effect are represented through the weights $a_1, a_2, ..., a_n$, meaning that the state of cause i will contribute a_i percent to the state of the effect. The weight vector $[a_1, a_2, ..., a_n]$ and the individual effect vectors for each cause are standardized, in a way that the sum of all the vector members is 1, in particular, $\sum_{i=1}^{n} a_i = 1$ and $\sum_{j=1}^{m} r_{ij} = 1, \forall i = 1, ..., n$.

With this model, the method in [12] allows to generate the effect's Conditional Probability Table (CPT) from the individual effect vectors and the weights through the following formula:

$$P(E = j \mid C_1 = j_1, C_2 = j_2, \ldots, C_n = j_n) = \sum_{i=1}^{n} a_i r_{i\sigma(j_i - j,j)} \tag{1}$$

in which $P(E = j \mid C_1 = j_1, C_2 = j_2, \ldots, C_n = j_n)$ is the conditional probability of the event in which the effect E has state j, while its causes $C_1, C_2, \ldots, C_n$ have state of $j_1, j_2, \ldots, j_n$ respectively; and $\sigma(j_i - j, j)$ is calculated as:

$$\sigma(j_i - j, j) = \begin{cases} j_i - j, & j_i - j \geq 0 \\ j, & j_i - j < 0 \end{cases}$$

$I(C_i)$, the influence of C_i to the effect, can also be calculated by formula (2), which is obtained from [12]. In the formula, $P(E = m \mid C_i = k)$ is the conditional probability when Effect E has state m (highest) and cause C_i has state k. By comparing the I value for each of the factors, one is able to identify which element is the most important.

$$I(C_i) = \frac{\left| \sum_{k=1}^{m-1} \frac{P(C_i = k)}{\sum_{j=1}^{m-1} P(C_i = j)} P(E = m \mid P(C_i = k)) - P(E = m \mid C_i = m) \right|}{P(E = m)} \tag{2}$$

2.3 Bayesian Network Approach to Transform the FAIR Framework

We consider the FAIR structure in Fig. 1 as a Bayesian network which consists of three pairs of cause-effect relations, including [cause: C, A; effect: TEF], [cause: Tcap, CS; effect: V], and [cause: TEF, V; effect: LEF]. Such cause-effect reasoning structure already forms a Bayesian network model. If one obtains the *individual effect vector* and the *cause weight* at the three nodes TEF, V, and LEF such as in Sect. 2.2, one can generate their CPTs. With the generated CPTs, this Bayesian model can generate statistical output for the LEF query, which can later be transformed into a numerical output. Next a method to identify the CPT at each node is proposed, e.g. between effect E and causes C_1, C_2, given their corresponding FAIR look-up table $[e_{ij} \in \{VL, L, M, H, VH\}, i = 1..5, j = 1..5]$ (refer to Fig. 2) and by means of the following steps:

Step 1. *Calculating the weight of the factors:* It is worth noting that the FAIR tables are formed based on the assumption that the states of the two causes create direct impacts on the state of the effect. Therefore, if one transforms the state data to numerical data, there should be a strong correlation between the causes and effect data in most of the cases. In the simplest form, one can assume the relation to be linear and translate the node state into a number by defining $VL = 1; L = 2; M = 3; H = 4; VH = 5$. One then has numerical data for the causes and effect, which can be used to run a regression to test the linear model between the causes and effect, $E = \alpha C_1 + \beta C_2 +$

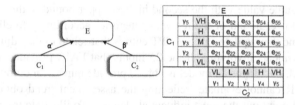

Fig. 2. Illustration of the cause-effect relation and the transformation parameters

ς (α and β are the coefficients and ς is the error). The coefficients α, β are then standardized with $\alpha' = |\alpha|/(|\alpha| + |\beta|)$ and $\beta' = |\beta|/(|\alpha| + |\beta|)$. We choose α' and β' as the weights of the causes toward the effect (see Fig. 2 below).

Step 2. Calculating the individual effect vector: In order to sharpen the difference between the levels of the state, one further converts state e_{ij} to n_{ij} in which $n_{ij} = k^{e_{ij}}$, $k > 0$. Therefore one has $n(VL) = k$, $n(L) = k^2$, $n(M) = k^3$, $n(H) = k^4$, and $n(VH) = k^5$. The weights are also set as $(\gamma_1, \gamma_2, \gamma_3, \gamma_4, \gamma_5)$ for the state of the cause (VL, L, M, H, VH), to further differentiate the effect of the state from the other causes (see Fig. 2). The choice of k and the state weights will not affect the correctness of the ranking, however, the larger the value, the deeper the numerical difference between the evaluation output of the threats. For each cause, one can derive its individual effect vector $r = [r(VL)$, $r(L)$, $r(M)$, $r(H)$, $r(VH)]$ by calculating the individual effect value of each state s_i as:

$$r(s_i) = \frac{\sum_{j=1}^{5} \gamma_i n_{ij}}{\sum_{l=1}^{5} \sum_{j=1}^{5} \gamma_l n_{lj}}, i = 1 \ldots 5$$

For each of the relations, and after obtaining the weight of the factors and the relevant individual effect vectors, one can generate the Bayesian CPT in each of the effect nodes following the formula in Sect. 2.2. Having the 3 CPTs from the 3 FAIR look-up tables is enough to form the overall Bayesian network for calculating the LEF output, given the input states of the causes.

Step 3. Generating numerical output: The output of the Bayesian model is a vector of the probability of the state evaluations for the LEF, for example, $[p_1, p_2, p_3, p_4, p_5]$, in which p_1 is the probability that LEF has state VL, p_2 is the probability that LEF has state L and so on. One uses the grade vector $[1, 2, 4, 8, 16]$ to derive the final numerical result (in detail the assessment for LEF is equal to $p_1 + 2*p_2 + 4*p_3 + 8*p_4 + 16*p_5$). This grade will be later used to compare and rank the threats, according to their LEF.

Step 4. Adjusting the Bayesian model for FAIR consistency: We also provide fix for inconsistences between FAIR and Bayesian model created by the weak correlation between values in the FAIR table. The fix will adjust the corresponding CPT entry of the Bayesian model based on the upper/lower bound based on the FAIR state. The 25 FAIR LEF outputs are grouped into 5 categories [VL L M H VH]. In each category, one replaces the FAIR output with the corresponding Bayesian grade (with the same input). The value range for each category is obtained next. If there is no intersection between the value ranges, the Bayesian model is fully consistent with the FAIR assessment. In case there are intersections, one decreases the upper bound (for instance, decrease to the same value with the second highest upper bound in the same category) or increase the lower bound of the relevant categories accordingly to eliminate all the intersections. One then updates all the CPT entries that relate to the adjustments. After this stage, assessments involving all the 25 inputs that FAIR provides are consistent.

Once formed, our Bayesian model is able to provide numerical output for the fuzzy inputs that FAIR cannot evaluate, reflecting the assessment trend obtained from the FAIR table, and point out the most influential element. To illustrate the method, in the next section the method is applied to a list of plausible threats to the smart grids.

3   Experimental Results and Discussion

3.1   Experimental Context and Input Data

In this section, we apply our LEF assessment for 14 threats (refer to the second column in Table 1), which are extracted from the 38 threats considering in our IRENE project [2]. The method to obtain the factor state is given in [2, 3]. Let us assume that after the evaluation, the inputs for the 14 threats are given in the third column of Table 1. Among the input, threat 9 and 13 have fuzzy values. This is because for threat 9, security experts were not able to agree to either assign the state "M" or state "H" to the "Tcap" factor. The chosen value indicates that a 40% value judgement was assigned to state "M" and a 60% to state "H". For threat 13, the experts were not able to evaluate the "A" factor at all, so an equal probability for each state was assigned. Although FAIR does not support assessments in these two cases, our method does handle such cases.

3.2   Results

Following Sect. 2.3 and choosing $k = 2$ and $(\gamma_1, \gamma_2, \gamma_3, \gamma_4, \gamma_5) = (1, 2, 3, 4, 5)$, the weights were obtained as follows [C A] = [0.39 0.61]; [Tcap CS] = [0.5 0.5]; [TEF V] = [0.7 0.3]. The resulting individual effect vectors are [C A] = [0.42, 0.34, 0.18, 0.05, 0.01; 0.49, 0.34, 0.13, 0.03, 0.01]; [Tcap CS] = [0.49, 0.3, 0.15, 0.05, 0.01; 0.01, 0.05, 0.15, 0.3, 0.49]; [TEF V] = [0.62, 0.25, 0.09, 0.03, 0.01; 0.37, 0.3, 0.23, 0.08, 0.02]. The Bayesian model is constructed using the formula in Sect. 2.2. LEF results for the threats calculated by this model are given in Table 1. To see how the change in the value judgement of fuzzy inputs can change the overall assessment of a threat, one varies it for the input of the "A" state for threat 13, while the other three factors [C, Tcap, CS] are fixed to [VL, L, VL]. The changes are represented in Fig. 3, in which different evaluation grades are calculated for an "A" input changing from [100%VL] to [20%VL 20%L 20%M 20%H 20%VH], [40%VL 15% L 15%M 15%H 15%VH], [60%VL 10%L 10%M 10%H 10%VH,…, and 100%VH]. The lower bound, which is the lowest value of the calculated set, is 264.49, and it results from "A" being at 100% "VL", while the upper bound is 531.3 when "A" is 100% "VH". The granularity of the evaluation can be observed in Fig. 3.

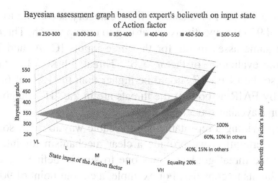

Fig. 3. Bayesian-FAIR evaluation of LEF with fuzzy state input in the "Action" factor

3.3 Discussion

Table 1 shows that a Bayesian network constructed on the basis of our method, generates assessments consistent with the FAIR framework. This is because the CPTs are
derived from the FAIR look-up tables and can be adjusted for ensuring consistency.
Moreover, our approach can differentiate further threats in the same category. For

Table 1. Numerical results of Bayesian-FAIR compared to FAIR.

ID	Name	Input state	FAIR	B-FAIR	Rank	MF(**)
1	Perimeter network scanning	[M, H, M, M]	H	889.5	7	C
2	Information gathering	[VH, H, M, H]	H	1016.9	4	C
3	Reconnaissance	[M, M, VL, L]	L	571.7	10	A
4	Craft phishing attacks	[H, H, VH, H]	H	1130.3	3	A
5	Spyware/Malware	[M, VH, H, VL]	VH	1147.1	1	A
6	Sniffers/Scanning	[M, H, H, M]	H	923.1	6	C
7	Insert subverted individuals	[M, H, H, VL]	H	939.9	5	CS
8	Exploit physical access	[L, M, L, H]	VL	342.9	13	A
9	Exploit unauthorized access	[H, M, 0.4 M–0.6H, VH]	n/a	290.33	14	A
10	Exploit split tunneling	[L, H, H, M]	M	685.1	9	C
11	Exploit mobile systems	[VH, H, VH, H]	VH	1147.1	1	C
12	Exploit recently vulnerabilities	[H, M, H, VH]	M	809.7	8	A
13	Physical compromise	[VL, E(*), L, VL]	n/a	343	12	A
14	Hardware compromise	[M, L, H, M]	VL	397.4	11	A

(*): State E indicates the equal probability of 20%VL – 20%L – 20%M – 20%H – 20%VH
(**): MF: Most influential factor

example, threats 6 and 7 are in the same "High" category according to FAIR, but have grades of 923.1 and 939.9 respectively according to our approach. The table shows that 6 and 7 have the same assessments for the three inputs [C, A, and Tcap], the only difference being the evaluation of factor "CS". Threat 7 has "VL" state compared to "M" for threat 6, so the LEF of 7 should be higher than the LEF of 6. This difference cannot be shown by FAIR as both of the threats are in the "H" category, but it can be seen clearly in our Bayesian model.

In addition to providing a repeatable and traceable way to reach some conclusions, even in case of uncertainties, we provide a clear mechanism for integrating a threat threshold. Having the threat grades allows one to simply define the cut out point to reduce the list of threats to consider. For example, a cut out point of 900 means threats are only considered when their grade is higher or equal to 900, reducing the list of threats to {2, 4, 5, 7, 6, 11}.

Our model is also able to handle fuzzy input. For example, for threats 9 and 13, the assessment grades of 290.33 and 343 respectively are given, while the FAIR model cannot provide the exact state. This capability is helpful when there is a lack of expert opinions for assessing the threats, or experts have conflicted assessments of the threats.

Another advantage is that our approach can point out the most influential factor for each of the threats. These outputs can be then combined to show which factor should be improved to lower the threat impact. For example, out of the 14 threats in Table 1, factor "A" is the one affecting the most threats (i.e. 8 out of 14). This suggests that system managers should implement countermeasures to lower the "Action", for example, by creating policies putting higher punishment on the attackers that initiated such threats, so as to lower attacker motivation. Such countermeasures will significantly lower the impacts of 8 out of 14 threats in the list, hence, effectively improving the security system with the least efforts for a particular smart grid configuration.

4   Conclusion

The ability to assess cyber threats is becoming more and more important for stakeholders, given the rise in smart grids. Due to the complexity of the risk assessment for complex systems such as urban smart grids, it is necessary to look for ways of considering large amounts of threats in a consistent manner and relate them to each other. In this paper, we proposed a method to transform the FAIR look-up tables to the Bayesian network model to provide a numerical threat LEF assessment combining elements of quantitative and qualitative methods. By applying the method to account for threats to a smart grid configuration, as shown in Sect. 3, we show that our method gives a consistent assessment with FAIR, while providing several more advantages, such as differentiating threats with the same FAIR inputs, giving more granular output, allowing flexible fuzzy inputs, and having the capability to highlight the most influential cause for particular threats, so as to effectively plan security countermeasures to lower the smart grid threats' impact. The interested reader can consult [10] for an elaborated example how the described approach can be used for considering countermeasures to several threats at once. In the future, we will extend this method for smart grid risk assessment.

Acknowledgments. This work was partially supported by the JPI Urban Europe initiative through the IRENE project.

References

1. Knapp, E.D., Samani, R.: Applied Cyber Security and the Smart Grid: Implementing Security Controls into the Modern Power Infrastructure. Elsevier Science, Burlington (2013)
2. IRENE, D2.1: threats identification and ranking (2015). http://www.ireneproject.eu
3. Jung, O., Besser, S., Ceccarelli, A., Zoppi, T., Vasenev, A., Montoya Morales, A.L., et al.: Towards a collaborative framework to improve urban grid resilience. In: Presented at the IEEE International Energy Conference, ENERGYCON 2016, Leuven, Belgium (2016)
4. NIST, Risk management guide for information technology systems (2002)
5. Farahmand, F., Navathe, S.B., Sharp, G.P., Enslow, P.H.: A management perspective on risk of security threats to information systems. Inf. Technol. Manage. **6**, 203–225 (2005)
6. Sun, L., Srivastava, R.P., Mock, T.J.: An information systems security risk assessment model under the Dempster-Shafer theory of belief functions. J. Manage. Inf. Syst. **22**, 109–142 (2006)
7. Peltier, T.R.: Information Security Risk Analysis. CRC Press, New York (2005)
8. Shameli-Sendi, A., Aghababaei-Barzegar, R., Cheriet, M.: Taxonomy of information security risk assessment (ISRA). Comput. Secur. **57**, 14–30 (2016)
9. Jones, J.: An introduction to factor analysis of information risk (fair). Norwich J. Inf. Assur. **2**, 67 (2006)
10. Vasenev, A., Montoya, L., Ceccarelli, A., Le, A., Ionita, D.: Threat navigator: grouping and ranking malicious external threats to current and future urban smart grids. In: Presented at the SmartGifts Conference on Smart Grid Inspired Future Technologies (2016)
11. RMI. FAIR basic risk assessment guide (2007). http://www.riskmanagementinsight.com/media/docs/FAIR_brag.pdf
12. Dui, H., Zhang, L.-L., Sun, S.-D., Si, S.-B.: The study of multi-objective decision method based on Bayesian network. In: 2010 IEEE 17th International Conference on Industrial Engineering and Engineering Management (IE&EM), pp. 694–698 (2010)

Replay Attack Impact on Advanced Metering Infrastructure (AMI)

Bashar Alohali[✉], Kashif Kifayat, Qi Shi, and William Hurst

School of Computing and Mathematical Sciences,
Liverpool John Moores University, Liverpool, UK
B.A.Alohali@2012.ljmu.ac.uk,
{K.Kifayat,Q.Shi,W.Hurst}@ljmu.ac.uk

Abstract. Advanced Metering Infrastructure (AMI) has currently become the most popular element in smart grid implementations both in home area network (HAN) and Neighborhood Area Network (NAN) environment as well as in large commercial/industrial establishments. The security of AMI has been an issue for several years, and many tools and utilities have been proposed to ensure the security of AMI networks. However, no network is completely safe from malicious users (hackers). Smart Meters (SM) in the NAN are typical targets for attackers, their objective being the acquisition of authentication information and attempting to successfully authenticate to become a part of the NAN. Such attacks are easy to launch and can cause significant impact since false data can be injected into the system. We explore the impact of such an attack on a previously developed authentication scheme and demonstrate that packet replays at a very fast rate can drain resources in a fashion similar to a Denial of Service (DoS) attack. The effect is pronounced since the authentication scheme uses a multi-hop path to reach the central authentication server. The intermediate nodes partially process each packet before forwarding it, causing an increase in the end-to-end delays as well as increased energy consumption. The authentication scheme is coded in C and the replay attacks are launched using an existing open source security tools.

Keywords: Smart grid · AMI · Replay attack

1 Introduction

Most of the existing world's electricity grid systems are built with centralised generation and one-way energy flows over the grid. Electricity is generated in central power plants and is distributed to different layers of customers over transmitting lines. However, the current electricity grid suffers many issues, such as limited protection, a limited control system, one-way communication, and limited participation with the customer [1]. The result is an incompetent and environmentally wasteful system unfit to be distributed, renewable solar and wind energy sources. The smart grid incorporates different information and communication (ICT) technologies for computer-based remote control and automation that are integrated to ensure effective electricity systems' delivery in the 21st century [13, 14].

© ICST Institute for Computer Sciences, Social Informatics and Telecommunications Engineering 2017
J. Hu et al. (Eds.): SmartGIFT 2016, LNICST 175, pp. 52–59, 2017.
DOI: 10.1007/978-3-319-47729-9_6

The recent cyber-attack on Ukraine's power grid has raised the security as significant requirement for the smart grid [2]. Increased security risk in the smart grid is a result of integration other technologies into the traditional power grid. The risks existing in other technologies are included along with the existing risks faced by the power grid. In addition, new risks stem from the integrated functionality of the smart grid, as is the case in SM. SMs use a microcontroller that has memory, digital ports, timers and real-time serial/wireless communication facilities [5].

This study makes two specific contributions. Firstly, it presents an analysis of the impact of the replay attack when the scheme uses multi-hop forwarding with the intermediate nodes decrypting the transit payload and re-encrypting it for a specific upstream node. The impact is illustrated in terms of the end-to-end authentication delays. Secondly, it proposes that for security schemes that involve multi-hop forwarding with authentication at every hop, a fairly high rate of replaying packets can lead to a denial-of-service like situation and such an attack be studied to mitigate its effect on the end-to-end network delays as well as the impact of node failures due to such attacks.

The rest of the paper is structured as follows. Section 1 briefly introduces the smart grid network model and the role of AMI in it. Section 2 discusses the security issues with AMI. Section 3 presents the related work. Section 4 discusses the Security scheme for AMI. Section 5 provides an overview of the impact of replay attacks on AMI; it then introduces the implementation of replay attack on AMI and analysis of replay attack over AMI and this section is followed by a conclusion.

1.1 Smart Grid Network

The Smart Grid (SG) is a heterogeneous network with multiple devices and technologies interconnecting them. As a data network, it comprises three parts – the Home Area Networks (HAN), the Neighbourhood Area Networks (NAN) and the Wide Area Network [4]. The devices in the HANs and NANs typically communicate with the control centre, which is accessible via the WAN. Interconnected end devices (sensor enabled appliances or sensors themselves) form the HAN. The HAN interconnects to the WAN via a SM, which is part of the NAN. Almost all the devices in the HAN and NAN are wireless communicating devices. The interconnectivity of SMs into NAN is collectively referred to as "Advanced Metering Infrastructure" (AMI) [8]. The focus of this paper is SMs in the NAN.

SMs are installed in domestic and commercial establishments, and have to be interconnected to communicate with upstream management entities [6]. They require forming a topology and the topology depends upon how they are distributed within a specific wireless range. Ideally, a single hop to the upstream node, typically performing 'gatewaying', 'security/authentication' and 'data aggregation' functions, is desired. This may be possible largely in dense areas such as structured high-rise buildings or malls. In a sparsely inhabited area or condominiums, the limited wireless range of the SMs might require a multiple hop path to the upstream node in the NAN. This implies that the intermediate nodes in the hop path must provide an authenticated forwarding of data from the downstream nodes. The security mechanisms deployed must ensure that each node is authenticated centrally as well as by the group. The focus of this paper is on a secure authentication scheme for such multi-hop networks formed by SMs in NAN. In the following section, we briefly discuss security in AMI.

2 AMI and Security

AMI uses wireless technologies in a public space. That makes it vulnerable to malicious users and poses security risks that range from meter tampering (physical) to online attacks such as Denial of Service, Replay, Sybil and so on [6]. Typically, the SM (and hence the AMI) implements a security scheme, which is built into their operational protocol. The core of such a scheme is the manner in which the secret keys are exchanged and other related cryptologic techniques used to ensure that the data from the originator to the recipient on the network is kept private and secure. Such a scheme is expected to mitigate malicious attacks. The impact of an attack, operationally, is not often quantified in terms of the impact on the end-to-end performance or as impact on the availability/resilience.

A replay attack involves a malicious node capturing authentication/data packets sent from a smart meter and re-sending them at a later point in time, expecting to authenticate and gain entry into the network [3]. There have been a great deal of research work on smart grid networks and [6–12] that propose a various authentication schemes that can prevent replay attacks. These schemes typically deploy a time stamp, a random number, a nonce value or a combination of these to prevent the attacks. However, to our knowledge, there is no literature that discusses the impact of a replay attack on end-to-end performance. Such impact studies exist for node capture attacks/node failures where the number of nodes affected for each node captured/failed in the network.

This paper makes an attempt to quantify the impact of a replay attack on the AMI, using a specific authentication scheme, in terms of the varying end-to-end delays and the amount of energy lost due to the attack. The receiving node will therefore receive a perfectly legal packet and the analysis of the data in the packet will enable the node to take a decision about the validity of the packet and its response. If the scheme is not sufficiently secure, the attacker node will be granted access to the AMI network with the privileges of other smart meters in the AMI.

The following section briefly details the security scheme under consideration and explains the replay attack scenarios, which are considered to check the impact on end-to-end performance.

3 Security Scheme

The security scheme considered uses group authentication. A group of smart meters (SMs) with one SM taking on the role of a gateway SM (GW) are interconnected in a manner that some SMs have a multi hop path to the group gateway (GW). The GW interconnects to the central authentication point, the Network operations center (NOC). SMs that are children of other SMs use the multi-hop path to reach the GW node. The

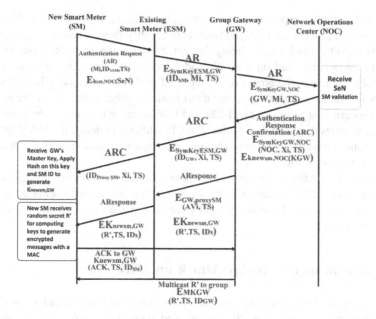

Fig. 1. The authentication process of a SM using a multi-hop path to the NOC

maximum hop count from the new SM to the GW SM is restricted to two. Packets travel from the new SM (L-SM), via an intermediate SM termed as the H-SM (proxy SM; ref Fig. 1) to the GW. The GW then forwards the packets to the NOC. All packets originating from the L-SM are encrypted. Data meant for the NOC is encrypted with the NOC's key and data meant for the GW is encrypted with the GW's key. The GW, therefore, decrypts every packet it receives and re-encrypts it using its shared key with the NOC. The reverse is done when the packets from the NOC go to the L-SM. The packet exchange for one authentication cycle is illustrated in the Fig. 1.

In the event of a replay attack, the attacker will resend a valid captured packet to the GW. Upon receipt, the GW will require to decrypt the packet using the shared key of the L-SM that it claims to arrive from. Following that, it forwards the packet to the NOC, which performs the same procedure and then implements the necessary checks on the content of the packet.

We have two specific concerns in the case of a replay attack. By definition, a replay attack involves resending a previously captured packet to gain access/privileges to the AMI. The security control is located at the central server, but the intermediate nodes have to process the packets they receive. The question then is whether the mitigation against this attack can occur at the intermediate nodes or whether it has to happen only at the central server. If the latter is true, then the intermediate nodes will needlessly be loaded. This is our primary concern. Secondly, an attacker can flood the network with the replay packets. This, in a sense, is a Denial of Service (DoS) attack, using replay packets. Such an attack can potentially cause higher damage to the intermediate nodes than a simple DoS attack. This is our second concern.

When the attacker repeatedly sends the replay packets, it causes the GW and the NOC to have to process the replayed packets to identify them. These packets are valid encrypted packets, which require being decrypted to examine the packet contents. This causes the processing load on the GW and NOC to increase. This increase can be substantial if the rate of the arrival of the replayed packets is sufficiently high, resulting in delays for traffic (authentication and data) from the other nodes, downstream.

Our concern is specifically on H-SM and GW nodes, which are in the path of the downstream nodes that send data to the NOC. In addition to delay, the H-SM and GW nodes consume energy to process the malicious packets and this could drain the resources on these nodes.

We attempt to estimate the impact of these two scenarios on the specific scheme mentioned above. We estimate this by implementing the security scheme and performing measurements of the delays and estimates of the energy consumption on a typical sensor mote.

4 Implementing the Replay Attack on AMI

In this section, we first present an analysis of the impact of replay attack on AMI, and then we explain the methodology that can be used to perform a replay attack over AMI.

4.1 The Impact of Replay Attacks

This section provides a quantifiable impact of replay attacks on AMIs. As mentioned earlier, replay attacks can be more harmful than Denial of Service attacks. This is because they can result in remote activities even though packets are encrypted. Replay attacks can alter authentication packets, allowing them to gain unauthorized access to the AMI. Once the attacker obtains access privilege to AMIs or smart meters, he/she can easily inject control indicators into the systems. The attacker has to initially study the packets being transferred from the customer's equipment to smart meters and examines these packets to identify the customer's general levels of power usage. Subsequently, such an attacker can spoof transmitted packets, and inject signals into the system.

To analyze the effect of replay attacks on AMI, we consider a scenario whereby there is a simple network topology. Say, a Sender-A has created either a 2, 3, or 4 hop (overlapping) transmission routes to a Receiver-B through relays R_1 and R_2. In such a situation, the attacker would be in Sender A's locality and eavesdrop on any packets being sent by A. Thus, as previously mentioned, the normal network route should include packets travelling from a group of smart meters through an intermediate SM, referred to H-SM, to the GW and the GW then forwards the packets to the NOC.

However, as seen above, the attacker can carry out any of the following activities:

1. The replay attacker can decide not to alter packet contents: Considering a situation where the PDR (i.e. the Packet Delivery Ratio) for all transmissions is '1'. If S sends a packet P_1 and R_1 receives this packet, followed by P_2, the attacker eavesdrop on these transmissions. Subsequently, R_1 forwards both P_1 and P_2 to R_2.

However, during intervals, the attacker can easily resend packet P_1 to R_1 again; thus, R_1 is misled and resends P_1 to R_2, resulting in a delay in the time taken to send both packets P_1 and P_2.

2. The attacker edits the packet header: The replay attacker can receive packet P_1, edit this packet, and then resend several of these packets to R_1, resulting in flooding of the network and higher time delay/discrepancy during transmission.

The sender-to-receiver performance degradation resulting from the actions of the replay attacker can be measured by the equation:

$$\Omega t_{\text{dynamic}}(\text{S D}) = T_S + \text{NAV C (TS)} + T_D \tag{1}$$

where:

T_S is the time to process message at Sender-S.

T_D is the time to process message at Receiver-D.

$NAV\ C(TS)$ is the time duration for communicating or sending packets between the sender (S) and receiver (D).

Using a more simplified analysis, we assume packets are sent from the smart meters to the gateway in "S_1" seconds, and from the gateway to the NOC in "S_2" seconds. In such a situation, a normal transmission from the sender to the receiver will take a maximum time interval of "$S_1 + S_2$" seconds. Considering a situation whereby a replay attacker eavesdrops on packets for "X_1" seconds, and then replays these packets for "X_2" seconds, the average time taken during a replay attack would be:

$$S_1^2 + X_2 + S_2^2 \tag{2}$$

Therefore, if an attacker listens during network packet transmissions for 100 s and then replays these packets for the next 100 s, applying the Eq. (2) above, time delay or discrepancy will be approximately 50% higher than the required time for sending packets. Furthermore, it can be deduced that when the hop count between nodes is increased (2 to 3 or 4), the time delay or discrepancy also increases. This is particularly based on the attacker's location, since the attack takes place from the location of the sender, the replayed messages travel through the major parts of the network with longer pathways, thereby resulting in increased time delays during transmission. Therefore, one replay attacker can reduce the routing time for packets by as much as 50%–60%, while numerous attackers can result in even more time disruption during network transmissions.

4.2 Methodology

We choose to perform the implementation of the secure scheme and the replay attack on Linux systems since there are several tools available and tested, for launching replay attacks. The implementation comprises of four programs done by a C program in a Linux machine, one each for the three SMs and the NOC, in Fig. 1. These programs communicate with each other using the packet content and packet format used by the secure scheme in Fig. 1. The interfaces on each of the machines were configured to transmit at 250 Kbps, the data rate of an ISM band ZigBee device. This emulates the link speeds, although it does not entirely model the radio link.

The authentication delay is first measured under normal conditions. Then, the L-SM is forced to re-authenticate while the replay attack is launched. The authentication time is measured when the attack is launched. The difference between the authentication times is plotted in Fig. 2. Note that the average increase is 10 ms with a rather large standard deviation of 5.8 ms. Note that these delay values are for this specific scheme. In this scenario, the increase in authentication time is accounted for by the processing times at the intermediate SMs.

In the authentication scheme in Fig. 1, the replay attacker replays the packet that the L-SM sends to the NOC. Upon receipt of the replayed packet at the intermediate SM, the packet is first decrypted, and subsequently the packet type and source are checked. Since the replayed packet is "genuine", the SM re-encrypts the packet and forwards the packet upstream. When the NOC detects it as a suspicious packet and drops it, there is no response sent back. However, the entities along the path to the NOC have to decrypt the packet and re-encrypt it before forwarding it one hop, upstream. It is this delay caused by the replay packet that manifests as the increase in the authentication time (10 ms, average) of the L-SM.

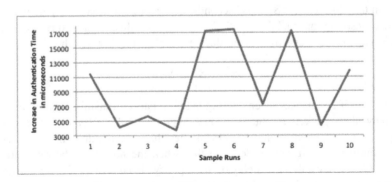

Fig. 2. Increase in authentication time due to a Replay attack

This result is only an initial result. The impact of the replay attack on the authentication time will increase with the increase in the number of L-SMs present as part of a H-SM's sub-tree (as the network scales) and likewise, the number of H-SMs that are part of a GW's sub-tree, in a cluster tree hierarchical topology. This is what we intend to measure, in addition to the impact of a DoS attack with replay packets. It is also possible to characterize "normal" operations by observing the inter-arrival time of packets, at each of the SMs. Such a characteristic can be applied as a template to detect replay attacks. This is currently, work-in-progress at our labs.

5   Conclusion and Future Work

Replay attacks pose a serious threat to smart grids and AMIs. Most security schemes for the AMI in the smart grid do not quantify the impact of an attack on the individual SMs in the AMI. In this paper we have attempted to analyse the impact of replay attacks on AMI. We have considered a case where the packets are replayed and quantified the impact of the attack in terms of increase in end-to-end authentication delay of the SM. Similar impact quantification for a larger network with more number of nodes is in progress. Along with this, the impact quantification of a DoS attack using replay packets is underway.

References

1. Berger, L.T., Iniewski, K.: Smart Grid Applications, Communications, and Security. Wiley, Hoboken (2012)
2. Tomkiw, L.: Russia-Ukraine Cyberattack Update: Security Company Links Moscow Hacker Group To Electricity Shut Down (2016, 8 Jan 2016). http://www.ibtimes.com/russia-ukraine-cyberattack-update-security-company-links-moscow-hacker-group-2256634
3. Alohali, B., Kifayat, K., Shi, Q., Hurst, W.: A survey on cryptography key management schemes for smart grid. J. Comput. Sci. Appl. **3**, 27–39 (2015)
4. Meng, W., Ma, R., Chen, H.-H.: Smart grid neighborhood area networks: a survey. IEEE Netw. **28**, 24–32 (2014)
5. Badra, M., Zeadally, S.: Key management solutions in the smart grid environment. In: Wireless and Mobile Networking Conference (WMNC), 2013 6th Joint IFIP, pp. 1–7 (2013)
6. Das, S., Ohba, Y., Kanda, M., Famolari, D., Das, S.K.: A key management framework for AMI networks in smart grid. IEEE Commun. Mag. **50**, 30–37 (2012)
7. Nian, L., Jinshan, C., Lin, Z., Jianhua, Z., Yanling, H.: A key management scheme for secure communications of advanced metering infrastructure in smart grid. IEEE Trans. Ind. Electron. **60**, 4746–4756 (2013)
8. Yee Wei, L., Palaniswami, M., Kounga, G., Lo, A.: WAKE: key management scheme for wide-area measurement systems in smart grid. IEEE Commun. Mag. **51**, 34–41 (2013)
9. Kim, J.-Y. Choi, H.-K.: An efficient and versatile key management protocol for secure smart grid communications. In: 2012 IEEE Wireless Communications and Networking Conference (WCNC), pp. 1823–1828 (2012)
10. Knapp, E.D., Samani, R.: Applied Cyber Security and the Smart Grid: Implementing Security Controls into the Modern Power Infrastructure. Elsevier Science, Burlington (2013)
11. Kamto, J., Lijun, Q., Fuller, J., Attia, J.: Light-weight key distribution and management for advanced metering infrastructure. In: 2011 IEEE GLOBECOM Workshops (GC Wkshps), pp. 1216–1220 (2011)
12. McLaughlin, S., Podkuiko, D., McDaniel, P.: Energy theft in the advanced metering infrastructure. In: Rome, E., Bloomfield, R. (eds.) CRITIS 2009. LNCS, vol. 6027, pp. 176–187. Springer, Heidelberg (2010). doi:10.1007/978-3-642-14379-3_15
13. Luo, L., Tai, N., Yang, G.: Wide-area protection research in the smart grid. Energy Procedia **16**(Part C), 1601–1606 (2012)
14. Gao, J., Xiao, Y., Liu, J., Liang, W., Chen, C.L.P.: A survey of communication/networking in Smart Grids. Future Gener. Comput. Syst. **28**, 391–404 (2012)

D²Sketch: Supporting Efficient Identification of Heavy Hitters Over Sliding Windows

Haina Tang[1], Yulei Wu[2(✉)], Tong Li[3], Hongbin Shi[3], and Jingguo Ge[3]

[1] School of Engineering Science, University of Chinese Academy of Sciences,
Beijing 100049, China
hntang@ucas.ac.cn
[2] School of Engineering, Mathematics and Physical Sciences,
University of Exeter, Exeter EX4 4QF, UK
y.l.wu@exeter.ac.uk
[3] Institute of Information Engineering, Chinese Academy of Science, Beijing 100195, China
{litong,shihongbin,gejingguo}@iie.ac.cn

Abstract. Heavy hitters can provide an important indicator for detecting abnormal network events. Most of existing algorithms for heavy hitter identification are implemented to deal with static datasets generated within a fixed time frame, lacking the ability to handle the latest arrivals of data streams adaptively. Considering the rigid demand for accurate and fast detection of outlier events in some networks like Smart Grids, these existing algorithms are not suitable to be deployed straightforward. To this end, this paper presents a new algorithm called D²Sketch for efficient heavy hitter identification over an adaptive sliding window for flexible dataset input. D²Sketch provides a novel framework that combines the Count-Min Sketch to get the connection degree of each host, with the stream-summary structure of Space Saving algorithm to get a more accurate list of Top-K heavy hitters. Moreover, it can adjust its measurement window to the most recent datasets automatically. Extensive experimental results show that the D²Sketch algorithm outperforms the related algorithm in terms of false positive rate, ordering deviation and estimate error.

Keywords: Heavy hitters · Count-min sketch · Space saving · Sliding window

1 Introduction

The network operators of Smart Grids are facing critical security challenges, because the infrastructure of Smart Grids is more vulnerable to cyber-attacks due to its distributed nature [1]. The existing researches revealed that cyber-attacks often result in a significant increase in network connections [2, 3]. For example, distributed denial-of-service (DDoS) attacks usually use plenty of geographically-distributed machines to send tremendous junk packets to the victim machine in order to deplete its resources and make the service unavailable to its legitimate users; to get these large number of machines being controlled, a worm infected host scans a large number of IP addresses in order to find vulnerable hosts to spread the worm.

© ICST Institute for Computer Sciences, Social Informatics and Telecommunications Engineering 2017
J. Hu et al. (Eds.): SmartGIFT 2016, LNICST 175, pp. 60–68, 2017.
DOI: 10.1007/978-3-319-47729-9_7

Heavy hitter refers to an element that occurs with a high frequency in the data streams. For network traffic, it usually represents a host connecting with a large number of IP addresses. Real-time identification of heavy hitters has been proved to be an efficient way to detect network anomalies like DDoS attacks, and, thus has led to a wide range of useful applications, such as click fraud detection and on-line analysis of stock market data.

Several solutions [4–6] have been reported in the current literature for the efficient identification of heavy hitters in traditional networks, e.g., virtual bitmap [5] and Space Saving [7]. Considering the high reliability requirement of Smart Grids, accurate and online detection of malicious and outlier events is in urgent need [8, 9]. On the other hand, existing algorithms for heavy hitter identification are not suitable to be deployed straightforward in Smart Grids due to the following problems. First, existing solutions are mainly implemented using static datasets such as NetFlow files collected in fixed time intervals. However, an abnormal event happening across two time intervals may be neglected for detection in this way. Second, due to the low scalability in high speed networks, the accuracy of existing algorithms [4, 10, 11] are not guaranteed during abnormal period when the number of flows in one time interval may increase sharply. Last, since existing operating systems are usually designed to monitor Top-K abnormal hosts or events, how to accurately find the Top-K heavy hitters with arbitrary ranking in one-pass on the data streams is necessary for Smart Grid network managers.

Targeting at those three challenges, we propose a new algorithm called D²Sketch to efficiently identify heavy hitters in networks like Smart Grids over an adaptive sliding window. Specifically, the proposed D²Sketch algorithm uses Count-Min Sketch to calculate the connection degree of each host, and adopts a revised and enhanced version of the Space Saving algorithm to get the TOP-K list of heavy hitters with arbitrary ranking. The most significant contribution of D²Sketch is that it implements a novel sliding scheme to dynamically adjust its measurement window in real-time. Extensive simulation experiments are conducted to validate the accuracy and evaluate the performance of the proposed algorithm. The results show that the D²Sketch outperforms the existing related algorithm in terms of false positive rate, ordering deviation and estimate error.

The remainder of the paper is organized as follows. After introducing the related work of heavy hitter identification in Sect. 2, we present the relevant definitions to be used in subsequent sections in Sect. 3. Section 4 elaborates the design of the proposed D²Sketch algorithm. Section 5 evaluates the performance of the algorithm via extensive simulation experiments, and, finally Sect. 6 concludes this work.

2   Related Work

Many existing works have been widely reported on identifying heavy hitters. For example, both Snort [10] and Flowscan [11] used hash tables to save all current flows within certain interval to find the heavy hitters. Although those techniques provide high accuracy on identifying heavy hitters, they are not scalable in high speed networks. Estan, Varghese, and Fisk [5] proposed a series of bitmap based algorithms to address this problem, where for each host the algorithm maintains a small bitmap to estimate the number of contacts during certain interval. Once the number of bits set in the small bitmap exceeds a certain threshold, a large multiresolution bitmap is allocated for that source to count the connections. This solution requires significant overheads to keep an index structure for mapping a source to its bitmap, resulting in a complicated process of maintaining bitmap for each IP address. The authors in [6] proposed a virtual indexing method which uses multiple hash functions to map the flow to a bitmap array for measuring the degree of host connections, but this mechanism may increase the probability of conflicts. Venkataraman et al. [4] proposed two flow sampling based techniques for detecting heavy hitters. Their one-level and two-level filtering schemes both use a traditional hash-based flow sampling technique for estimating connection degrees. When this scheme is adopted for high speed connections, its estimation accuracy decreases. Nuno Homem et al. [14] presents an FSW (Filtered Space-Saving with Sliding Window) algorithm to address this problem which is not a lightweight solution and requires to maintain a histogram for each counter in bitmap. Recently, Zhen et al. [13] designed a BF_LRU which adopts LRU to evict mice flows and Bloom filters to conserve heavy hitters in fixed intervals, however, it could not provide the sliding windows solution as considered in our proposed D^2Sketch.

3 D^2Sketch: TOP-K Heavy Hitter Identification Algorithm

3.1   Establish the Data Structure

As shown in Fig. 1, our proposed algorithm—D^2Sketch—consists of two kinds of data structures. The first part is a count-min sketch [12] which is fast and compact in a very simple way, and the second part is the Stream-Summary structure of space saving algorithm. The connection degree of each host in data streams is calculated using count-min

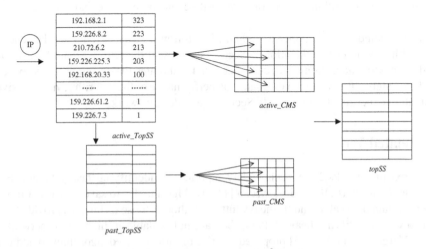

Fig. 1. The data structure of D^2Sketch

sketch and provided as the input of space saving which greatly decreases the over-estimation errors in space saving algorithm.

D^{2}Sketch deals with the datasets received in the latest 2s elements which are divided into two sub_windows. For IP network, 2s can be initialized as the expected maximum flows per second normally. So the sliding scheme of D^{2}Sketch is implemented using double count-min sketch and three stream-summary structures. The datasets for the latest s elements ($[(p-1) \times s, p \times s](p \geq 2)$) are mapped to *active_topSS* and *active_CMS*, whereas *past_topSS* and *past_CMS* represent datasets for the previous s elements ($[(p-2) \times s, (p-1) \times s](p \geq 2)$). And *topSS* saves the final TOP-K heavy hitters for the set of the latest 2s elements in data streams.

active_topSS or *past_topSS* is utilized to keep the TOP-K elements sorted by their estimated connection degrees which is a Stream-Summary data structure. It can be denoted as an array with m length, and each entry in this array maintains a 2-tuple (ip, c), where ip represents the IP address, and c denotes the counter of its connection degree in the set of current or previous s elements.

active_CMS or *past_CMS* is a count-min sketch with an array of counters of width w and depth d. Each incoming item is mapped uniformly into the range $\{1, 2, \ldots w\}$ using d hash functions ($h_1, h_2 \ldots h_d$), then the value of the corresponding cell is updated. Elements not existing in *active_topSS* or *past_topSS* are mapped to *active_CMS or past_CMS*, separately.

3.2   Update the Connection Degree for Each Host in the Latest s Elements

Let $S = s_1, s_2 \ldots s_n$ be the set of input data stream that arrives sequentially in the latest s elements. Each element s_i can be briefly expressed as ($sourceIP_i, destinationIP_i$). Let $CDegree_{active}(sourceIP_i)$ represent the connection degree of $sourceIP_i$ and *minfq* represent the minimum counter value in *active_topSS*.

For the new flow s_i denoted by ($sourceIP_i, destinationIP_i$), D^{2}Sketch algorithm will check whether it is in *active_topSS*. If so, suppose the index for $sourceIP_i$ is s. Then *active_topSS*[s] [1] is updated by adding 1. And the connection degree of $sourceIP_i$ is calculated as following:

$$CDegree_{active}(SourceIP_i) = active\_topSS[s][1] \tag{1}$$

If $sourceIP_i$ is not in *active_topSS* and *active_topSS* is not full, $sourceIP_i$ will be inserted into the tail of *active_topSS*.

$$CDegree_{active}(SourceIP_i) = 1 \tag{2}$$

Otherwise, D^{2}Sketch computes the $CDegree_{active}(SourceIP_i)$ from *active_CMS*, and compares $CDegree_{active}(SourceIP_i)$ with *minfg*.

$$CDegree_{active}(SourceIP_i) = \min(active\_CMS[j, h_j(SourceIP_i)] + 1), 1 \leq j \leq d \tag{3}$$

If $CDegree_{active}(SourceIP_i)$ is greater than *minfg*, the item at the tail of *active_topSS* is removed from *active_topSS* and added to *active_CMS*. Then, it would insert

($SourceIP_i$, $CDegree_{active}(SourceIP_i)$) into *active_topSS*. Otherwise, D^2Sketch will update the *active_CMS* as follows:

$$active\_CMS[j, h_j(SourceIP_i)]+ = 1, 1 \leq j \leq d \tag{4}$$

3.3   Calculate the TOP-*K* Heavy Hitter in the Latest 2 *s* Elements

We use an $m \times n$ bit array, *topSS[i][j]* ($0 \leq i < m, 0 \leq j < 2$), to store the TOP-*K* heavy hitters in the latest 2*s* elements, and keep a value *mintop* as the minimum counter value in *topSS*. Upon receiving a new flow s_i denoted by ($sourceIP_i$, $destinationIP_i$), D^2Sketch can get the connection degree of $sourceIP_i$ in the latest 2*s* elements based on the method used in Sect. 3.1 for the calculation of $CDegree_{active}(SourceIP_i)$. At the same time, D^2Sketch calculates the connection degree of $sourceIP_i$ in the previous *s* elements as $CDegree_{past}(SourceIP_i)$ from *past_topSS* and *past_CMS* in the same way. Then the connection degree of $sourceIP_i$ in the latest 2*s* elements can be expressed as:

$$CDegree(SourceIP_i) = CDegree_{active}(SourceIP_i) + CDegree_{past}(SourceIP_i) \tag{5}$$

If $SourceIP_i$ has be included in *topSS* with index *s'*, then:

$$topSS[s'][1]+ = CDegree_{active}(SourceIP_i) \tag{6}$$

Otherwise, if $CDegree(SourceIP_i)$ is greater than *mintop*, the item with the counter equal to *mintop* is deleted from *topSS*, and ($SourceIP_i$, $CDegree(SourceIP_i)$) is inserted into *topSS*.

3.4   Output the TOP-*K* Results in the Dataset of Last 2 *s* Elements

At the end of each sliding window for the set of past $[(p-2) \times s, p \times s](p \geq 2)$ elements in data streams, D^2Sketch will output the TOP-*K* results from *topSS* directly. Then, the following operation is executed to prepare for the next sliding window of last $[(p-1) \times s, (p + 1) \times s]$ elements:

$$topSS, past\_topSS = active\_topSS; \tag{7}$$

$$past\_CMS = active\_CMS; \tag{8}$$

$$Initialize(active\_topSS); Initialize(active\_CMS); \tag{9}$$

4   Complexity Analysis

D^2Sketch is designed to minimize the false positive rate of heavy hitter identification over sliding window, so we do not consider space complexity, but focus on time complexity. There are three operations in the worst case within each iteration in D^2Sketch for processing a newly arrived element *e*: scanning *active_topSS, past_topSS* and *topSS* to search for the existence of *e*, probing *d* cells to calculate the connection degree of *e* in *active_CMS* and *past_CMS*, updating the same *d* cells as probed in the last step by adding 1 in *active_CMS*. According to [13], the time cost per update of SpaceSaving with simple heap implementations is $O(\log k)$. Suppose all these operations are time-equal, thus, the amortized time complexity of each iteration is $O(3\log k + 3d)$. Since the access to *active_topSS*, *past_topSS*, *active_CMS*, and *past_CMS* can be executed in parallel, the amortized time complexity of each iteration can be $O(2\log k + 2d)$.

5   Performance Evaluation and Experiments

In this section, we present experimental results that we have conducted to verify the the effectiveness of our proposed algorithm for identifying heavy hitters. To show its merits, the performance of D^{2}Sketch is compared with that of Space Saving (SS) algorithm [14] which is a popular algorithm in this field and is the most related algorithm to our design. We use a real traffic trace collected at the International Link of CSTNet (China Science & Technology Network) to evaluate the performance of the algorithms, where 288 traffic trace files were collected on 1st Oct, 2015 with one flow file generated every 5 min.

5.1   Evaluation Metrics

We define the following metrics to evaluate the performance of our algorithm.

1. The false positive rate (f.p.r): The false positives (f.p) refer to those hosts who are actually not in top-*k* heavy hitter list, but they are estimated to be in the final *topSS* list. The false positive rate is defined as f.p.r : = |f.p|/*k*.
2. Ordering deviation: The false ordering instance refers to those hosts whose OrderID is different from their estimated OrderID. The ordering deviation means the number of false ordering instances in the final *topSS* list.
3. Estimate error: The frequency deviation of a host is defined as the absolute value of its actual frequency subtracting its estimated frequency. The estimate error means the summation of the frequency deviation of all hosts in the final *topSS* list.

5.2   Experimental Results

We compared D^{2}Sketch and SS algorithms on 288 NetFlow trace sample files. For each file, we use D^{2}Sketch and SS to estimate the TOP_*K* IP-pair and their frequency, respectively; in addition, we calculate the real value with traditional sum and sort algorithms. Based on that data, we calculate the average false positive rate, ordering deviation and frequency deviation sum from Top-10 to Top-100 to demonstrate the accuracy of the proposed D^{2}Sketch algorithms.

Figure 2 demonstrated the ability of D^{2}Sketch to find the correct TOP-*K* heavy hitters set. To achieve a fair comparison among all the candidate algorithms, our experimental results are based on the 288 trace files, and the *K* is to be 10, 20, 30, 40, 50, 60, 70, 80, 90, and 100, separately. For each point on the figure, we conducted our experiment using

288 trace sample file, and calculate the average result as the experimental result. Clearly, our proposed algorithm is able to correctly identify the TOP-K heavy hitters with remarkable accuracy. For example, the false positive rate for D^{2}Sketch is almost always below 0.56 % while it reaches 5.23 % for SS algorithm when $K = 100$.

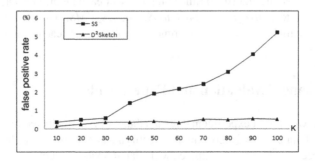

Fig. 2. False positive rate of D^{2}Sketch and SS

To highlight the ability of our algorithm to calculate the corresponding distinct values for TOP_K heavy hitters with higher accuracy, we evaluate the ordering deviation (Fig. 3) and estimate error (Fig. 4) of D^{2}Sketch algorithm compared with that of SS. The ordering deviation and the estimate error of D^{2}Sketch keep almost half of that derived from SS algorithm. The false positive rate of D^{2}Sketch keeps almost horizontal but SS increases sharply. Thus, the experimental results validate our approach, demonstrating that our proposed algorithm is viable and effective solution for detecting heavy hitters in large ISP networks and Smart Grids with various key applications, e.g., DDoS detection.

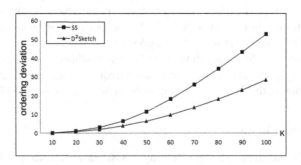

Fig. 3. Ordering deviation of D^{2}Sketch and SS

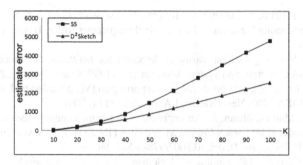

Fig. 4. Estimate error of D²Sketch and SS

6 Conclusion

In this paper, we have developed an efficient approach called D²Sketch to identify heavy hitters over sliding windows which mainly consists of two parts: *count-min sketch* is used to get the connection degree of each host on the network, and *space saving* is redesigned and enhanced to get a more accurate Top-*K* heavy hitter list. In addition, we have designed new evaluation metrics to compare the performance of our proposed algorithm with that of space saving algorithm. The extensive experimental results have indicated that our algorithm noticeably outperforms the existing popular space saving algorithm in terms of false positive rate, ordering deviation and estimate error. The results have suggested that the proposed algorithms are of great value for heavy hitter identification in Smart Grids.

References

1. Goel, S.: Anonymity vs. security: the right balance for the smart grid. Commun. Assoc. Inf. Syst. **36**(1) (2015). Article 2
2. Zhao, Q., Kumar, A., Xu, J.: Joint data streaming and sampling techniques for detection of super sources and destinations. In: IMC. ACM Press, Berkeley, pp. 77–90 (2005)
3. Kompella, R.R., Singh, S., Varghese, G.: On scalable attack detection in the network. In: Proceedings of the 4th ACM SIGCOMM Conference on Internet Measurement, pp. 187–200 (2004)
4. Venkataraman, S., Song, D., Gibbons, P.B., Blum, A.: New streaming algorithms for fast detection of superspreaders. In: Proceedings of the 12th ISOC Symposium on Network and Distributed Systems Security (SNDSS), pp. 149–166 (2005)
5. Estan, C., Varghese, G., Fisk, M.: Bitmap algorithms for counting active flows on high speed links. In: ACM SIGCOMM Internet Measurement Workshop (2003)
6. Wang, P., Guan, X., Gong, W., Towsley., D.F.: A new virtual indexing method for measuring host connection degrees. In: INFOCOM 2011, pp. 156–160 (2011)
7. Metwally, A., Agrawal, D., Abbadi, A.E.: Efficient computation of frequent and Top-k elements in data streams. In: Proceedings of 10th International Conference on Database Theory (ICDT 2005), pp. 398–412 (2005)
8. Liu, J., Xiao, Y., Li, S., et al.: Cyber security and privacy issues in smart grids. IEEE Commun. Surv. Tutorials **14**(4), 981–997 (2012)

9. Marques, C., Ribeiro, M., Duque, C., Ribeiro, P., Da Silva, E.A.B.: A controlled filtering method for estimating harmonics of off-nominal frequencies. IEEE Trans. Smart Grid 3(1), 38–49 (2012)
10. Roesch, M.: Snort–lightweight intrusion detection for networks. In: Proceedings of the USENIX LISA Conference on System Administration 1999, Seattle, WA, pp. 229–238 (1999)
11. Plonka, D.: Flowscan: a network traffic flow reporting and visualization tool. In: Proceedings of USENIX LISA 2000, New Orleans, LA, pp. 305–317 (2000)
12. Cormode, G., Muthukrishnan, S.: An improved data stream summary: the count-min sketch and its applications. In: Farach-Colton, M. (ed.) LATIN 2004. LNCS, vol. 2976, pp. 29–38. Springer, Heidelberg (2004). doi:10.1007/978-3-540-24698-5_7
13. Homem, N., Carvalho, J.P.: Finding top-k elements in a time-sliding window. Evolving Syst. 2(1), 51–70 (2011)
14. Zhang, Z., Wang, B., Lan, J.: Identifying elephant flows in internet backbone traffic with bloom filters and LRU. Comput. Commun. 61, 70–78 (2015)
15. Cormode, G., Hadjieleftheriou, M.: Methods for finding frequent items in data streams. VLDB J. 19(1), 3–20 (2010)

Short Term Load Forecasting
for Residential Buildings
An Evaluation Based on Publicly Available Datasets

Carola Gerwig[(✉)]

University of Hildesheim, 31141 Hildesheim, Germany
gerwig@uni-hildesheim.de

Abstract. Short Term Load Forecasting is an essential component for optimizing the energy management of individual houses or small micro grids. By learning consumption patterns on smart metering data, smart grid applications such as Demand-Side-Management can be applied. However, most of the research done in this field is based on data which is not publicly available. Moreover, the evaluations also vary in the evaluation settings and the error measurements. In this work, five state-of-the-art approaches are compared on three publicly available datasets in the most common scenarios. By doing this, the most promising methods and model settings are pointed out. Furthermore, it can be seen that forecasting the consumption 24 h ahead achieves about the same accuracy as doing it four hours ahead. Still, the best results for individual households are rather inaccurate. By aggregating ten households, the results enhance by a factor of about 60%.

Keywords: Short term load forecasting · Energy management · Intelligent buildings · Smart grid · Machine learning

1 Introduction

The integration of renewable energy resources in the existing energy systems and the digitalization of the energy sector offer great potential for implementing smart energy management applications. Thereby, accurate short term load forecasting (STLF) can significantly improve the micro-balancing capabilities of the energy systems [1]. On the basis of reliable forecasts, load shifting methods can be implemented to cushion peaks in demand and supply. With the increasing number of smart meters in residential buildings[1] smart grid applications can also be applied to the residential level. Therefore, specific short term load forecasting methods are required as the variance of the load is extremely high and established methods can fail easily.

To apply STLF to residential buildings, suitable methods have to be developed and evaluated. The standard methods for STLF such as stochastic time

---

[1] For example, in Germany the installation of smart meters in new buildings has been enforced since 2010 by law, cf. 21b Abs. 3a EnWG.

© ICST Institute for Computer Sciences, Social Informatics and Telecommunications Engineering 2017
J. Hu et al. (Eds.): SmartGIFT 2016, LNICST 175, pp. 69–78, 2017.
DOI: 10.1007/978-3-319-47729-9_8

series, artificial neural networks and others have been applied and adapted. Unfortunately, the presented evaluation is often done on datasets which were recorded in individual research projects and are not publicly available. But the results strongly depend on the size of the demand and the consumption behavior of the residents. As a result, the methods and results are often not comparable.

Given the importance of STLF to balance energy supply and demand, more research is needed. In this work, five state-of-the-art methods are implemented and compared on three publicly available datasets. In the experiments forecasting horizons of 1 h and 24 h are considered. The examined methods are an autoregressive model, k-nearest neighbors, decision trees, random forests and a kernel ridge regression as well as a persistent forecast and an averaging method as benchmarks. The accuracy is measured with three standard error measurements. Relevant features to improve the forecasting accuracy are introduced in settings of the experiments and discussed in the results.

The explorative approach of this article leads to suggestions about future research directions. This includes not only the choice of methods but also the choice of the size of the training data and the input parameters. Furthermore, it is pointed out, that forecasts for *several* households – even for a small number such as ten – achieve significantly better results. This leads to the consideration whether smart grid applications should primarily be adapted not for individual households, but for residential blocks.

The paper is organized as follows: Sect. 2 gives an overview of the related work. In Sect. 3, the datasets, the forecasting methods and the settings of the experiments are described. In Sect. 4, the experimental results are presented and the conclusions are reached in Sect. 5.

2   Related Work

Articles about STLF for residential buildings are mainly from the last five years as the necessary Smart Meter technology for residential buildings was introduced to the market in around 2010. Frequently used methods are linear regression (cf. [2–5]), more advanced autoregressive methods (cf. [6–8]) and artificial neural networks (cf. [2,3,9]). Furthermore, clustering methods are proposed: In [4,10] similar time sequences are matched while [11] concentrates on a customer classification. A newer approach are Support Vector machines in this context. They are applied in [2,4,8]. Exponential smoothing and Kalman filters are further options applied in a few articles. An extensive literature review of the state-of-the-art on STLF can be found in [12].

The results of the evaluations carried out in the named articles have contradictory results. In [6,8] autoregressive methods achieve better results than neuronal networks for individual users. Both articles recommend aggregating the load of more than 20 households as the error variance of one household is too high for good results. Contrary to this outcome, [7] states that an neuronal network performs slightly better than the autoregressive method used. Linear regression outperforms a Multi Layer Perceptron and SVR for fewer than 32 households in two other articles, [2,4]. Exponential Smoothing does not perform well on individual households according to [13].

In [12] it is pointed out that comparing the results of the different papers is difficult as the evaluations do vary not only in datasets but also in the length of the time horizon which is forecast, in the granularity of the sampling dataset and the forecast, and in the choice of error measures. Therefore, the results differ considerably. The MAPE varies between 7.3% [9] and 49% [7] for individual households. Only a few articles [2,4,7,11] provide evaluations on publicly available datasets so that the results of most articles found are not reproducible and cannot be used for benchmarking.

3   Experiments

3.1   Data

Due to the increasing attention paid to smart metering, several publicly available datasets for energy consumption have been published recently. For the experiments in this articles, datasets are chosen which have been used for the evaluation of STLF methods before. These are the *Reference Energy Disaggregation Dataset* [14] (REDD) and the *CER Electricity Dataset*[2] (CER). Additionally, the *Almanac of Minutely Power dataset* [15] (AMPds) is used as it is a comprehensive dataset and provides measurements for almost two years. The datasets contain readings in different granularities. In order to compare the results, they are transformed into time series with hourly resolution.

REDD. It is provided by the Massachusetts Institute and contains the consumption data of six households for 18 days in the Spring 2011. It has been used in [7]. Following the data selection carried out in that article, the low frequency readings of the individual appliances of house no. 1 are aggregated and the longest consistent distinct dataset used.

CER. It provides the half-hourly demand of almost 5000 Irish homes and businesses from 2009 to 2010. It is also used in [2,4,11]. It is not clear, which datasets have been used in the mentioned articles; therefore, ten datasets which do not have missing values and whose measurements cover a similar time range are chosen. These are the recordings of the houses 1000, 1003, 1004, 1005, 1006, 1009, 1013, 1014, 1015 and 1018. The common time range comprises 535 days.

The forecasting methods are applied to each of these houses individually; the average of the individual error measures is taken as accuracy measurement. This evaluation is referred as **CER_ind**. Additionally, the forecasting methods are applied to the aggregation of the datasets. This dataset is referred as **CER_agg**.

AMPds. It includes the energy consumption data of one house in Vancouver for almost two years (728 days).

---

[2] http://www.ucd.ie/issda/data/commissionforenergyregulationcer/.

3.2 Forecasting Methods

For the evaluation, an autoregressive model (**AR**) and four machine-learning methods are chosen. These are k-nearest neighbor regression (**KNN**), Decision Trees (**DT**), Random Forest Regression (**RF**), kernel ridge regression (**KRR**). Two simple benchmarks are used: a persistent forecast (**PER**), in which the forecast values are equal to the last observation, and an averaging method (**AVG**), in which the forecast values are the average of the training data for the specific time of day. For all methods, a sliding window strategy is applied, this means the training data is a sliding window of specific lengths in the time series and the model is refitted for every forecast. The four machine learning methods are implemented by using Scikit-learn in Python for machine learning methods [16].

The forecasting method are evaluated for a forecasting horizon of one and 24 h. The forecast of 24 h is done by separately forecasting the consumption k hours ahead for $k \in \{1, \ldots, 24\}$.

3.3 Input Parameters

Typical input parameters for STLF methods are the perceding values of the time series as well as information about the time of day and the type of day. For each method, several settings are evaluated for the STLF for one hour by doing a grid search over the combination of the following input parameters. For the STLF of 24 h, a reduced grid search is applied by using only the three best settings obtained by the STLF for an hour for each method.

Autocorrelated Preceding Values. Examining the values of the partial autocorrelation function, one finds a strong autocorrelation between the directly preceding values and the value 24 h ago.

Using this indication, the following lags are tested as input for the STLF for 1 h: Let x_{t+1} be the consumption value at time $(t + 1)$ that should be forecast, then $\{x_t, x_{t-1}\}$, $\{x_t, x_{t-23}\}$, $\{x_t, x_{t-1}, x_{t-23}, x_{t-24}\}$ and $\{x_t, x_{t-1}, x_{t-23}, x_{t-167}\}$ are used as possible input parameters for all methods except the kernel ridge regression. KRR can handle a great number of features before it overfits. Therefore, the input parameters are set to $\{x_t, x_{t-1}, x_{t-2}, x_{t-23}, x_{t-24}\}$, $\{x_t, x_{t-1}, x_{t-23}, x_{t-167}\}$ or to all precedings values 24 h back, i.e. $\{x_t, \ldots, x_{t-23}\}$.

The REDD dataset is a very small set to learn consumption patterns, therefore forecasts which only use the preceding value x_t as input are tried out additionally for this dataset.

Naturally, these input parameters have to be shifted for a forecast of 24 h as the directly preceding values are not given. Let x_{t+k} be the value that should be forecast at time t for $k \in \{1, \ldots, 24\}$. To predict it, $\{x_t, x_{t+k-24}\}$, $\{x_t, x_{t-1}, x_{t+k-24}, x_{t+k-25}\}$ or $\{x_t, x_{t-1}, x_{t+k-24}, x_{t+k-168}\}$ are used.

Average of Preceding Values. In order to capture seasonal patterns, the average of the previous day and the average of the previous week are added as optional input parameters.

Time of Day. Many energy-intensive activities such as cooking depend on the time of day. Therefore, the time of day is added as input parameter. This input parameter needs to be scaled as the KNN approach uses the Euclidean distance to find neighbors. Therefore, it is normalized with respect to the preceding consumption values and multiplied with one third of the standard deviation of the training data. These input parameter are not used in the AR approach as it builds 24 different regression models – one for each hour of the day.

Type of Day. To capture the consumption patterns which are specific for a certain weekday, the information about the weekday can also be used as input parameter (0 for Monday, 1 for Tuesday, ...). Three variations are tested: (1) without information about the weekday, (2) different values for each day of the week, (3) different values for Saturday, Sunday and Monday and the same values for the rest of the week. In the same way as before, this input parameter is scaled by multiplying it with the standard deviation of the training data.

3.4 Size of Training Data

The size of the training data is quite a performance factor in machine learning methods – generally speaking the bigger the better. But as some consumption patterns depend on the season of the year, two different settings are evaluated: (1) 91 days (i.e. a quarter of the year) and (2) a large dataset consisting of four fifths of the original dataset as training data. The evaluation is done on the remaining days. Only for the short REDD dataset only seven days are applied – this is the longest size of training data used in [7].

3.5 Error Measurements

The most popular error measurement to evaluate forecasts is the mean average percentage error (MAPE). Applied to STLF for residential buildings, the MAPE has one major drawback: In periods, in which the consumption is nearly zero, the MAPE becomes extremely large. Therefore, the normalized mean absolute error (NMAE) and the normalized root mean square error (NRMSE) are evaluated, too:

$$\text{MAPE} = \frac{1}{n}\sum_{t=1}^{n}\left|\frac{x_t - \hat{x}_t}{x_t}\right|, \ \text{NMAE} = \frac{\sum_{t=1}^{n}|x_t - \hat{x}_t|}{\sum_{t=1}^{n}|x_t|}, \ \text{NRMSE} = \sqrt{\frac{\sum_{t=1}^{n}(x_t - \hat{x}_t)^2}{\sum_{t=1}^{n}x_t^2}},$$

where x_t is the actual value at time t and $\hat{x}_t$ is the forecast one (cf. [4]).

To look into every evaluation done on the individual houses for the CER dataset would go beyond the scope of this article, therefore the average of the

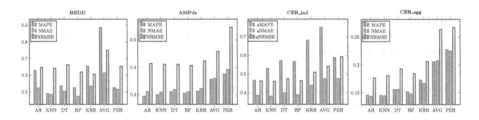

Fig. 1. Accuracy errors for STLF for one hour.

results for CER_ind is presented. The same applies for the forecasts for 24 h. Thereby, the average over the results for the single hours for each method is taken.

4   Experimental Results

4.1   STLF for One Hour

The results differ considerably on the four datasets. The exact results are shown in Fig. 2 and visualized in Fig. 1. The next section highlights some observations.

First of all, the results on the REDD dataset are rather poor; the advanced methods hardly beat the persistent forecasts. The MAPE is between 49% and 66%. This corresponds to the results of [7], in which the MAPE for this setting is between 51% and 63%. A training set of only seven seems to be too small.

The results on the AMPds dataset are quite similar for all methods. AR, KNN and RF perform slightly better than the other methods. This observation also applies to the results on CER_ind except for the MAPE. It is relatively high in comparison to the other error measurements. Analyzing the individual results, the main factor for this are the high MAPE values of house 1000 and 1006. For house 1000 this might be due the fact that the consumption is very low for some days in the test sets. For house 1006 one can see a strong dependency on the season of the year. These patterns are captured best by the AR method.

Aggregating the data of ten houses has a great influence on the errors for the CER data. The errors decrease by a factor of almost 70% for the MAPE and more than 60% for the NMAE and the NRMSE. Again, the best results are achieved for AR and KNN.

4.2   STLF for 24 Hours

The results of the STLF for 24 h are presented in Fig. 3. As before, the accuracy of the predictions for the REDD dataset is quite low. Nevertheless, compared to the results of [7] for sliding windows, KNN outperforms the presented approaches in the article by 20% concerning the MAPE.

The best results on the AMPds and CER datasets are again achieved by AR, KNN and RF. These methods are quite competitive. It seems, that KNN

	REDD			AMPds			CER_ind			CER_agg		
	MAPE	NMAE	NRMSE	MAPE	NMAE	NRMSE	aMAPE	aNMAE	aNRMSE	MAPE	NMAE	NRMSE
AR	.6305	.5240	.6471	**.2945**	.3111	.4145	**.4664**	.3861	.4630	.1461	**.1437**	**.1765**
KNN	**.4899**	.4852	.6440	.2997	**.3090**	.4113	.5310	**.3817**	**.4612**	**.1453**	.1450	.1807
DT	.5355	.5049	.6640	.3108	.3186	.4111	.5716	.3992	.4751	.1557	.1563	.1928
RF	.5254	**.4747**	.6197	.3053	.3105	**.4071**	.5668	.3890	.4636	.1519	.1481	.1842
KRR	.6564	.5356	**.6065**	.3144	.3223	.4236	.6564	.5356	**.6065**	.1726	.1671	.2057
AVG	.8875	.6137	.7501	.3560	.3596	.4581	.8875	.6137	.7501	.2056	.2075	.2630
PER	.5415	.5252	.6550	.3750	.3923	.5456	.5415	.5252	.6550	.2271	.2246	.2664

Fig. 2. Accuray errors for one hour-STLF.

	REDD			AMPds			CER_ind			CER_agg		
	aMAPE	aNMAE	aNRMSE	aMAPE	aNMAE	aNRMSE	aMAPE	aNMAE	aNRMSE	aMAPE	aNMAE	aNRMSE
AR	1.0798	0.7947	0.9938	**.3325**	**.3323**	**.4358**	.6970	**.4449**	**.5106**	.1709	**.1601**	.2008
KNN	**0.7047**	**0.5843**	**0.7014**	**.3330**	.3340	.4367	**.6883**	.4471	.5136	**.1673**	.1620	**.1993**
DT	0.7126	0.6260	0.7435	.3428	.3436	.4505	.7106	.4595	.5243	.1833	.1748	.2141
RF	1.3293	0.9145	0.8981	.3461	.3342	**.4352**	.7235	.4483	**.5100**	.1745	.1656	.2019
KRR	0.7840	0.6760	0.8122	.3457	.3481	.4513	.8364	.5071	.5647	.2060	.1946	.2345
AVG	0.8875	0.6137	0.7501	.3560	.3596	.4581	.7496	.4748	.5420	.2056	.2075	.2630
PER	1.2282	0.9004	1.1650	.7524	.6300	.7351	1.647	.8537	.8990	.7556	.6043	.6563

Fig. 3. Average accuracy errors for 24 hour-STLF.

obtains the best MAPE, AR the best NMAE and RF achieves the best results with respect to the NRMSE. Again, the results for the aggregated CER datasets are far more promising than for CER_ind.

For all methods and datasets, the errors increase slightly for the first four hours and stay almost stable afterwards (cf. Fig. 4). The average results for AMPds and CER_agg are only 2–3% worse than the results for STLF for one hour. However, the enhancements in comparison to the averaging method decrease.

4.3   Analysis of the Experiment Settings

In this subsection, the experiment settings for AMPds and CER_agg are examined in order to find guidelines for the input parameters.

Concerning the length of the training data, the NRMSE is lowest for all methods on both datasets, when the bigger training set is used. For AMPds the absolute difference to the small training set is about 0.5%, for CER_agg about 2%. This also applies for the MAPE and NMAE on CER_agg. On AMPds, they get better (about 0.5%) when only 91 days are used as training data.

The best selection of preceding values also depends on the choice of the error measurement for AMPds. Using the preceding values $\{x_t, x_{t-1}, x_{t-23}, x_{t-167}\}$ delivers the best results with respect to the MAPE and NMAE, while concerning the NRMSE $\{x_t, x_{t-1}\}$ is preferable for all methods except AR and KRR. For CER_agg it is $\{x_t, x_{t-1}, x_{t-23}, x_{t-167}\}$ independent of the choice of error measurement for all methods except DT, which works best with $\{x_t, x_{t-1}\}$.

Using the average values of preceding daily or weekly consumption is not useful concerning the MAPE on both datasets. But concerning the NMAE and

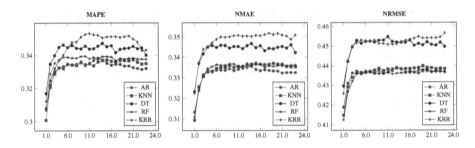

Fig. 4. Evaluation results for the single hours of the STLF for 24 h for AMPds.

the NRMSE the average consumption of the previous day has to be a proven be useful parameter for AR and KNN on AMPds and for all methods on CER_agg. Additionally using the average consumption of the previous week improves the NRMSE for AR and KNN on AMPds and for RF and DT on CER_agg.

The analysis of the input options for the type of the day differs on the two datasets; on AMPds AR works best without it. For KNN, using the distinction between working days and weekends works best. The results of DT and RF also enhance by this distinction. Furthermore, when using the big training set of 580 days, it can be recommended to use the differentiated consideration of all weekdays for these two methods. For CER_agg AR and KNN work best by differentiating between working days and weekends, while DT, RF and KRR profit from the differentiated consideration of all weekdays.

5   Conclusions

Five regression methods for STLF for 1 h and 24 h have been evaluated on three different consumption datasets. The results vary between the datasets and it can only be recommended to always use several sets for the evaluation of STLF methods.

On the whole, the autoregressive approach and the k-nearest neighbor method achieve the best results. However, Random Forests are quite competitive for the AMPds dataset. The kernel ridge regression doesn't seem to be a good choice for STLF. It can be assumed, that by tuning the feature selection small enhancements for all methods could be achieved.

The analysis done on the settings of the experiments points out, that the size of the training data is of great importance. The experiments on the REDD dataset with only seven days for training deliver poor results. Although comparable results for 91 days and 580 days as size of the training data have been achieved for AMPds, the results of CER_agg suggest that increasing the training data generally improves the results and in particular the NRMSE.

The results of the STLF for 24 h indicate that the predictions 24 h ahead are only slightly worse than the forecast four hours ahead. Comparing it to STLF of one hour there is a drop in accuracy of 2–3%.

Overall, the achievable accuracy for STLF for individual households seems to be quite low. The best results for single households so far have a mean average percentage error of about 30%. By aggregating ten houses, the MAPE can be reduced substantially. For the CER dataset it drops down to 17%, which is an improvement by a factor of almost 70% on this dataset. Therefore, depending on the succeeding usage of the forecasts, it should be considered whether it is possible to integrate several households in the STLF for the desired usage.

References

1. E-Energy Abschlussbericht: Ergebnisse und Erkenntnisse aus der Evaluation der sechs Leuchtturmprojekte (2013). http://www.e-energie.info/documents/ E-Energy_Ergebnisbericht_Handlungsempfehlungen_BAUM_140212.pdf
2. Humeau, S., Wijaya, T., Vasirani, M., Aberer, K.: Electricity load forecasting for residential customers: exploiting aggregation and correlation between households. In: Sustainable Internet and ICT for Sustainability, SustainIT (2013)
3. Javed, F., Arshad, N., Wallin, F., Vassileva, I., Dahlquist, E.: Forecasting for demand response in smart grids: an analysis on use of anthropologic and structural data and short term multiple loads forecasting. Appl. Energy **96**, 150–160 (2012). Elsevier
4. Wijaya, T., Humeau, S., Vasirani, M., Aberer, K.: Residential Electricity Load Forecasting: Evaluation of Individual and Aggregate Forecasts (2014)
5. Iwafune, Y., Yagita, Y., Ikegami, T., Ogimoto, K.: Short-term forecasting of residential building load for distributed energy management. In: Energy Conference (ENERGYCON), pp. 1197–1204 (2014)
6. Tidemann, A., Høverstad, B., Langseth, H., Öztürk, P.: Effects of scale on load prediction algorithms. In: IET (2013)
7. Veit, A., Goebe, C., Tidke, R., Doblander, C., Jacobsen, H.: Household Electricity Demand Forecasting-Benchmarking State-of-the-Art Methods. arXiv preprint arXiv:1404.0200 (2014)
8. Sevlian, R., Rajagopal, R.: Short Term Electricity Load Forecasting on Varying Levels of Aggregation. arXiv preprint arXiv:1404.0058 (2014)
9. Yang, H., Liao, J., Lin, C.: A load forecasting method for HEMS applications. In: 2013 IEEE Grenoble PowerTech (POWERTECH), pp. 1–6 (2013)
10. Fujimoto, Y., Hayash, Y.: Pattern sequence-based energy demand forecast using photovoltaic energy records. In: Renewable Energy Research and Applications (ICRERA), pp. 1–6 (2012)
11. Chaouch, M.: Clustering-based improvement of nonparametric functional time series forecasting: application to intra-day household-level load curves. IEEE Trans. Smart Grid **5**(1), 411–419 (2014)
12. Gerwig, C.: Short term load forecasting for residential buildings—an extensive literature review. In: Neves-Silva, R., Jain, L.C., Howlett, R.J. (eds.) Intelligent Decision Technologies. SIST, vol. 39, pp. 181–193. Springer, Heidelberg (2015). doi:10.1007/978-3-319-19857-6_17
13. Rossi, M., Brunelli, D.: Electricity demand forecasting of single residential units. In: 2013 IEEE Workshop Environmental Energy and Structural Monitoring Systems (EESMS), pp. 1–6 (2013)

14. Kolter, J., Johnson, M.: REDD: a public data set for energy disaggregation research. In: Workshop on Data Mining Applications in Sustainability (SIGKDD), San Diego, CA (2011)
15. Makonin, S., Popowich, F., Bartram, L., Gill, B., Bajic, I.: AMPds: a public dataset for load disaggregation and eco-feedback research. In: Electrical Power & Energy Conference (EPEC), pp. 1–6. IEEE (2013)
16. Pedregosa, F., Varoquaux, G., Gramfort, A., Michel, V., Thirion, B., Grisel, O., Blondel, M., Prettenhofer, P., Weiss, R., Dubourg, V., Vanderplas, J., Passos, A., Cournapeau, D., Brucher, M., Perrot, M., Duchesnay, E.: Scikit-learn: machine learning in python. J. Mach. Learn. Res. **12**, 2825–2830 (2011)

Short-Term Electrical Load Forecasting Based on Fuzzy Logic Control and Improved Back Propagation Algorithm

Lei Wan[1], Junxiu Liu[1(✉)], Senhui Qiu[1,2], Mingcan Cen[1], and Yuling Luo[1(✉)]

[1] Guangxi Key Lab of Multi-Source Information Mining and Security,
Faculty of Electronic Engineering, Guangxi Normal University, Guilin 541004, China
`vanley_lei@139.com,`
`{liujunxiu,qiusenhui,mingcancen,yuling0616}@gxnu.edu.cn`
[2] Guangxi Experiment Center of Information Science, Guilin 541004, China

Abstract. The short-term electrical load forecasting plays a significant role in the management of power system supply for countries and regions. A new model which combines the fuzzy logic control with an improved back propagation algorithm (FLC-IBP) is proposed in this paper to improve the accuracy of the short-term load forecasting (STLF). Specifically, the composite-error-function-based method and the dynamic learning rate approach are designed to achieve a better predictable result, which mainly applies the improved back propagation algorithm (BP). Besides, the fuzzy logic control theory is used to build up a good optimization process. Experimental results demonstrate that the proposed method can improve the accuracy of load prediction.

Keywords: Short-term load forecasting · Back propagation · The fuzzy logic control theory

1 Introduction

The STLF is of great significance for the plan, scheduling and operation of the power system. Accurate load forecasting can improve the quality of electric energy, reduce the operation cost and increase the security and stability of the system by making full use of network managing efficiency [1]. A variety of approaches have been implemented to forecast the load demand. According to the techniques employed, these approaches can be generally classified into two categories: the classical prediction method and the modern prediction method [2]. The classical approaches for the load forecasting, such as time series method [3], regression models [4] etc., are based on similarity in forecasting of future power load curve by using the foregone information [5]. In general, though these techniques are effective for the forecasting of short-term load on normal days, they always fail to yield good results when special events occur [6]. Then, the modern approaches of load forecasting are based on the grey forecasting model [6], artificial neural networks, and some hybrid forecasting methods. These forecasting methods showed a better accuracy performance to a certain extent. However, compared to the medium and long-term load forecasting, the STLF is so easily influenced by certain random factors that it can cause load trend instability, so there is still some key points and difficulty for load forecasting to

© ICST Institute for Computer Sciences, Social Informatics and Telecommunications Engineering 2017
J. Hu et al. (Eds.): SmartGIFT 2016, LNICST 175, pp. 79–86, 2017.
DOI: 10.1007/978-3-319-47729-9_9

considerate [7]. Inspired by the hybrid forecasting model, a novel method combining fuzzy logic control with improved back propagation (FLC-IBP) is designed for STLF in this paper, which can enhance the accuracy of load prediction. Simulation results demonstrate that the proposed method is feasible and effective.

2 Related Work

According to the characteristics of STLF, neural network has received increasing attention as a non-linear and dynamic system with fault tolerance and strong robustness [8]. This section summarizes reported efficient forecasting approaches based on neural network method.

The implementation of a spiking neural network for STLF model was proposed in [9] to forecast one day ahead and one week ahead hourly demand pattern. Another STLF model based on BP learning algorithm was proposed in [7], which was trained by the artificial immune system learning algorithm. In order to achieve a better accuracy performance, some hybrid forecasting models are presented, which combined at least one model by analyzing the characteristics, advantages and disadvantages of these models and they are at present the main research area for load forecasting [10]. The key topic is about utilization of combination forecasting method. So a new BP algorithm based on genetic algorithm model (GA-BP) was proposed in [8], which is a typical combined architecture. The GA-BP model applied genetic algorithm to optimize the weights and threshold of BP artificial neural network for STLF. Similar to [8], a neuro-fuzzy model was implemented in [11], and a computational intelligent technique was implemented to optimize artificial neural network architecture in [12]. Inspired by the above hybrid forecasting models, a FLC-IBP model in this paper is proposed which aims to achieve a better accuracy performance via combining fuzzy logic control with improved back propagation.

3 FLC-IBP Algorithm

3.1 Improved BP Algorithm

As for the traditional BP algorithm, the sum of the squares of the error is usually used as the objective function, and the gradient descent method is applied to deal with the minimum value of error function. However, the curved surface of the error function is often irregular. As a result, it is difficult for traditional gradient descent method to handle the irregular curved surface, that is, some sample points will make gradient be close to zero in the curved surface, which may make the convergence rate of network become slow and make the training fall into a standstill state as well even though there is not a local minimizer in this situation. Besides, as the curved surface of error function may have multiple local minimizers in terms of convex and non-convex functions, there will be a great impact for the training precision of BP algorithm, which will make network reach the local minimizer rather than the global optimal point. Therefore, on the one hand, the design of error function is very important. On the other hand, the static learning rate of traditional BP algorithm based the gradient descent method should be improved to increase the convergence rate of training.

Then, a composite error function strategy based on error rate is proposed to optimize the learning process of BP algorithm, which separates the error rate of hidden layers and the error rate of output layer. The detailed present is as follows,

$$E_{new}(n) = \gamma E(n) + (1 - \gamma)E_{hidden}(n) \tag{1}$$

where n represents the training steps, $E(n)$ is the error function of the traditional BP algorithm and γ is the error rate, which is given by:

$$\gamma = \frac{o(n) - t(n)}{t(n)} \tag{2}$$

Specifically, $o(n)$ is the output value of the training steps and $t(n)$ is the objective value of samples in the training steps. Besides, the $E_{hidden}(n)$ in Eq. (1) represents the error function in the hidden-layer. It is given by Eq. (3), where y_{pk} is the output of the hidden-layer, 0.5 is the desired output, and p and k are the layer and node number of the network, respectively.

$$E_{hidden}(n) = \frac{1}{2} \sum_{p=1}^{P} \left(y_{pk} - 0.5 \right)^2 \tag{3}$$

The log-sig function is usually used as the active function of the hidden-layer in the traditional BP algorithm. According to the log-sig function, the medium value of log-sig function is 0.5; and when the output of log-sig function tends to 0 or 1, its output value tends to infinity. When y_{pk} tends to 0.5, the value of $E_{hidden}(n)$ reaches the local minimizer. Then, according to Eqs. (1) and (3), the original error function $E(n)$ is still used, thus the convergence rate of the network is kept.

At the first stage of training, there is a difference between the target value of the sample and the actual output value, which can make γ value very large. Then, we use the original error function $E(n)$. After several training steps, the value of γ may become smaller and smaller. Then the user-defined error function $(1 - \gamma)E_{hidden}(n)$ is used to avoid the BP algorithm training falling into a local minimum.

In accordance of the above analysis, the error function of the original BP algorithm and user-defined hidden-layer error function are merged to be a composite error function, and the error rate is taken as weights. According to weights, we can judge the current training stage of network. When there is a faster convergence rate for network training, the original error function is used; otherwise, the user-defined hidden-layer error function will be adopted to adjust weights with a larger step for escaping from local minimums and accelerating the convergence rate.

The weight modulation formula of the traditional BP algorithm is shown as Eq. (4), where the learning rate η is a static parameter, and the hidden-layers and output-layer have the same weight modulation formula [13]. Therefore, it is difficult to adjust the weight and decrease the convergence or divergence speed.

$$\Delta W_{kj} = -\eta(t_k - o_k)f'(y_k) \cdot \frac{\partial y_k}{\partial w_{kj}} = -\eta(t_k - o_k)f'(y_k) \cdot x_k \tag{4}$$

On the contrary, due to the dynamic η and the consideration of the hidden-layers' training, the improved BP algorithm can self-adopt the learning rate η in the terms of error. The weight adjustment formulas of output-layer and hidden-layer are shown as follows,

$$\Delta W_{kj} = -\eta_2(t_k - o_k)f'(y_k) \cdot \frac{\partial y_k}{\partial w_{kj}} = -\eta_2(t_k - o_k)f'(y_k) \cdot x_k \tag{5}$$

$$\Delta v_{ji} = \eta_1 \sum_{p=1}^{P} \left(-\frac{\partial E_p}{\partial Net_{pj}} \cdot \frac{\partial Net_{pj}}{\partial v_{ji}} \right) = \eta_1 \sum_{p=1}^{P} \left(\sum_{k=1}^{m} \delta_{pk}w_{kj} \right) y_{pj}(1 - y_{pj})X_{pi} \tag{6}$$

where η_1 and η_2 are given as follows respectively:

$$n_1 = \begin{cases} \eta_1(n-1) - \alpha\left(\dfrac{Max(\Delta E(n-1), \Delta E(n-2))}{E(n)}\right), & if \Delta E(n-1) \ and \ \Delta(n-2) \le 0 \\ \eta_1(n-1) - \alpha\dfrac{\Delta E(n-1)}{E(n)}, & if \Delta E(n-1) > 0 \\ \eta_1(n-1), else \end{cases} \tag{7}$$

$$n_2 = \begin{cases} \eta_2(n-1) - \alpha\left(\dfrac{Max(\Delta E(n-1), \Delta E(n-2))}{E(n)}\right), & if \ \Delta E(n-1) \ and \ \Delta(n-2) \le 0 \\ \eta_2(n-1) - \alpha\dfrac{\Delta E(n-1)}{E(n)}, & if \Delta E(n-1) > 0 \\ \eta_2(n-1), else \end{cases} \tag{8}$$

where the function of the interval $\alpha \in (0.01, 0.03)$ is similar with a filter to prevent learning rate from excessive volatility to ensure the stability of the algorithm.

On the one hand, the error value of the network is relatively larger if $\Delta E > 0$, which means that this network is likely to join the steep region of the curved surface to excess the weight adjustment. Then, there will need to reduce the adjustment of ΔW_{kj} and Δv_{ji} to achieve a balance of convergence rate. As shown in Eqs. (5), (6), (7) and (8), the values of learning rates η_1 and η_2 can be decreased to adjust weight values ΔW_{kj}, Δv_{ji} and convergence speed of network in the improved algorithm. On the other hand, if $\Delta E(n-1)$ and $\Delta(n-2) \le 0$, the network joins in the flat region, so there will need to increase the adjustment of weight values. Therefore, the values of learning rate η_1 and η_2 can be increased to adjust the weight. In addition, if $\Delta E < 0$, the training error will decrease, then there only need to keep learning rate unchanged.

3.2   FLC-IBP Algorithm

The term fuzzy logic was introduced in the fuzzy set theory by Zadeh, L.A. [14]. Fuzzy logic has been applied to many fields, from control theory to artificial intelligence. In this work, the corresponding models based on fuzzy logic control theory are built to describe some fuzzy, random data or parameters (i.e. weather, temperature, special events, etc.).

FLC-IBP is just based on this fuzzy logic control, and its flow diagram is shown in Fig. 1. Specifically, some input data is directly transmitted to the IBP algorithm to learn, and the other is used to judge in the fuzzy logic controller according to knowledge base, fuzzy rules and membership function. Once the fuzzy reasoning judgment is finished, the fuzzy logic controller would feedback the results to IBP algorithm to learn. Then, the controlled object realizes the output, calculates the error between actual output and the ideal output, and further feedback item to IBP algorithm and fuzzy logic controller to make the IBP neural network continue to adjust the connection weights and thresholds.

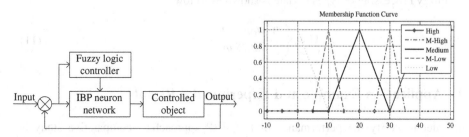

Fig. 1. The working principle of FLC-IBP   **Fig. 2.** The membership functions curve of temperature

3.3   The Establishment of Short-Term Power Load Forecasting Model Based on Proposed FLC-IBP

An improved BP algorithm and the related knowledge of FLC-IBP are presented in Sects. 3.1 and 3.2 The FLC-IBP algorithm combines the advantages of the artificial neural network and fuzzy logic control, where artificial neural network is responsible for pattern recognition and adaptive adjustment, while reasoning and making decision is completed by fuzzy logic control.

In order to increase the prediction accuracy, the factors that affect the power load forecasting, such as weather, temperature, etc. should be taken into account as well. The following part mainly analyzes the fuzzy processing of these random factors.

(1) *Fuzzy processing of temperature*

There are five membership functions for fuzzy processing of temperature. Considering the simplicity of the calculation, the trapezoidal and triangular membership functions in this work are used as Fig. 2.

Let temperature T_h denote the above five curves, low temperature, medium-low temperature, medium temperature, medium-high temperature and high temperature's membership degree are respectively shown as $\{T_{h1}, T_{h2}, T_{h3}, T_{h4}, T_{h5}\}$. According to the maximum value of membership degree $T_{h\text{-}max} = \max \{T_{h1}, T_{h2}, T_{h3}, T_{h4}, T_{h5}\}$, we can obtain the membership function of T_h.

(2) *Fuzzy Processing of weather conditions*

According to the annual weather condition of the forecast area, the fuzzy processing of weather is shown as the following (Table 1):

Table 1. The fuzzy processing of weather

Weather	Heavy snow	Moderate snow	Light snow	Heavy rain	Moderate rain
Value	0.1	0.2	0.3	0.4	0.5
Weather	Light rain	Cloudy day	Cloudy	Sunny	
Value	0.6	0.7	0.8	0.9	

(3) *Fuzzy Processing of types of date*

Fuzzy Processing of types of date is shown as follows:

$$U = \begin{cases} 0, Mon. - Fri. \\ 1, Sat., Sun. \end{cases} \tag{11}$$

4   Application Example and Experimental Results

In order to verify the performance of FLC-IBP approach, a comparative study is conducted by comparing the results with the prediction obtained from the traditional BP learning algorithm. The historical load data set from March 28th to May 1st of 2011 in Eastern China is chosen to validate the model, and the data is divided into two parts, that is, one training part is between March 28th and April 24th, the other is between April 25th and May 1st.

The results during training process as followed: the FLC-IBP algorithm is converged at the 193-rd iteration, while the BP learning algorithm takes 380 iterations to converge; the time of FLC-IBP to learn the behavior of the data during the training process is about 8 s shorter than that of BP algorithm. Further, the true values and forecasting values proceeded by BP algorithm and the proposed method (FLC-IBP) on the week-days and the weekends are shown in Figs. 3 and 4, where two different case studies based on mapping of input data are investigated. They are uncorrelation and similar moment correlation [9]. *Uncorrelation:* the loads data on the same day are only processed by the normalization method for training, the correlation between different dates are very weak.

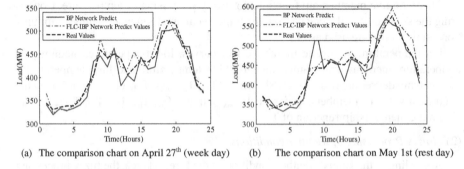

(a) The comparison chart on April 27th (week day)   (b)   The comparison chart on May 1st (rest day)

Fig. 3. The real value and predicted value comparison chart with uncorrelated data input on week day and rest day

Similar moment correlation: the similar moments on different days are correlated for training i.e. that, 12:00 of the previous day is correlated to 12:00 of the present day.

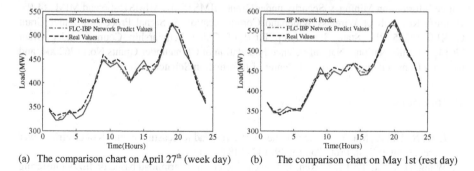

(a)   The comparison chart on April 27th (week day)   (b)   The comparison chart on May 1st (rest day)

Fig. 4. The real value and predicted value comparison chart based on data processing method of similar moment on week day and rest day

Obviously, the result from two different case studies is that similar moment correlation gives a better forecasting accuracy for week-days and weekends; The mean absolute percentage error MAPE [7] for a week is shown in the Table 2. Then, the average value of MAPE of FLC-IBP during week days is calculated to equal to 3.38 %, and BP algorithm is equal to 5.09 %. Similarly, as for the other days, the FLC-IBP MAPE is 3.59 %, and BP algorithm is 6.76 %. The value of MAPE is smaller, which indicates a better prediction. From these results, it can be shown that the FLC-IBP approach can reach a more accurate forecasting for the load curves than BP model.

Table 2.   The mean absolute percentage error of FLC-IBP and BP

Date	BP MAPE (%)	FLC-IBP MAPE (%)
2011-4-25 (Monday)	6.51	4.53
2011-4-26 (Tuesday)	4.28	2.29
2011-4-27 (Wednesday)	4.78	3.23
2011-4-28 (Thursday)	5.33	4.52
2011-4-29 (Friday)	4.56	2.33
2011-4-30 (Saturday)	5.66	3.31
2011-5-1 (Sunday)	7.89	3.87

5   Conclusions

The composite error function and the dynamic learning rate method were proposed in this paper to improve BP algorithm. Besides, this paper combined fuzzy logic control theory to build a better optimization algorithm (FLC-IBP). The prediction capability of FLC-IBP was tested by using the historical load data set during the period of March 28th to May 1st, 2011. And compared with BP learning algorithm, the experimental results showed that the proposed FLC-IBP algorithm has a better forecasting accuracy.

Acknowledgements. This research was partially supported by the Guangxi Natural Science Foundation under Grant 2015GXNSFBA139256 and 2014GXNSFBA118271, the Research Project of Guangxi University of China under Grant ZD2014022, Guangxi Key Lab of Multi-source Information Mining & Security under Grant MIMS15-07, MIMS15-06 and MIMS14-04, Guangxi Key Lab of Wireless Wideband Communication & Signal Processing under Grant GXKL0614205, the State Scholarship Fund of China Scholarship Council under Grant [2014]3012, the National Natural Science Foundation of China under Grants No.11262004, and the grant from Guangxi Experiment Center of Information Science.

References

1. Liao, N., Hu, Z., Ma, Y.: Review of the short-term load forecasting methods of electric power system. Power Syst. Prot. Control **39**, 147–148 (2011)
2. Wang, Q., Zhou, B., Li, Z., Ren, J.: Forecasting of short-term load based on fuzzy clustering and improved BP algorithm. In: International Conference on Electrical and Control Engineering, pp. 4519–4522 (2011)
3. Hagan, M.T., Behr, S.M.: The time series approach to short term load forecasting. IEEE Trans. Power Syst. **2**(3), 785–791 (1987)
4. Thomas, S.D.: Comparative models for electrical load forecasting. Eur. J. Oper. Res. **26**, 176–177 (1986)
5. Ferreira, L., Andersson, T., Imparato, C., Miller, T., Pang, C., Svoboda, A., Vojdani, A.: Short-term resource scheduling in multi-area hydrothermal power systems. Int. J. Electr. Power Energy Syst. **11**, 200–212 (1989)
6. Li, W., Han, Z.: Application of improved grey prediction model for power load forecasting. In: 12th International Conference on Computer Supported Cooperative Work in Design, pp. 1116–1121 (2008)
7. Hamid, M.B.A., Rahman, T.K.A.: Short term load forecasting using an artificial neural network trained by artificial immune system learning algorithm. In: 12th International Conference on Computer Modelling and Simulation, pp. 408–413 (2010)
8. Wang, Y., Ojleska, V., Jing, Y., Tatjana, K., Dimirovski, G.M.: Short term load forecasting: A dynamic neural network based genetic algorithm optimization. In: 14th International Power Electronics and Motion Control Conference, pp. 157–161 (2010)
9. Kulkarni, S., Simon, S.P.: A new spike based neural network for short-term electrical load forecasting. In: 4th International Conference on Computational Intelligence and Communication Networks, pp. 804–808 (2012)
10. Kumar, S., Ranjeeta, D., Dash, B.P.K.: A hybrid functional link dynamic neural network and evolutionary unscented Kalman filter for short-term electricity price forecasting. Neural Comput. Appl. **10**, 1–18 (2015)
11. Koushki, A.R., Nosrati Maralloo, M., Lucas, C., Kalhor, A.: Application of neuro-fuzzy models in short term electricity load forecast. In: 14th International CSI Computer Conference, pp. 41–46 (2009)
12. ul Islam, B., Baharudin, Z., Raza, M.Q., Nallagownden, P.: Optimization of neural network architecture using genetic algorithm for load forecasting. In: 5th International Conference on Intelligent and Advanced Systems, pp. 1–6 (2014)
13. Bai, S., Zhou, X., Xu, F.: Spectrum prediction based on improved-back-propagation neural networks. In: 11th International Conference on Natural Computation. pp. 1006–1011 (2015)
14. Zadeh, L.A.: Fuzzy sets. Inf. Control **8**, 338–353 (1965)

On the Study of Secrecy Capacity
with Outdated CSI

Tong Chen[1(✉)], Peng Xu[2], Zhiguo Ding[1], and Xuchu Dai[2]

[1] Lancaster University, Lancaster, UK
t.chen2@lancaster.ac.uk
[2] University of Science and Technology of China, Hefei, China

Abstract. In this paper, we study the secrecy capacity of the proposed transmission scheme. It is provided that the SNR at the selected best relay, destination and eavesdropper are same due to the symmetrical setup. The outage probability is considered here to evaluate the performance of such a secrecy communication system. Then the system will be extended into outdated channel state information (CSI) scenarios. The correlation coefficients between the actual and outdated channel will influence the system performance. The closed-form expressions of the achievable outage probability will be derived to demonstrate the performance of the proposed secure transmission scheme, and numerical results are presented.

Keywords: Secrecy communication · Outage probability · Outdated CSI

1 Introduction

The study of privacy and security in wireless communication networks has played an important role with widespread use of wireless networks. The key idea of information theoretic secrecy is to ensure the uncertainty at the eavesdropper is larger than the transmitted data rate which avoids being intercepted by eavesdroppers [1–3]. In wireless communication scenarios, because of the dynamic feature of wireless channels making eavesdroppers unavoidable, it is difficult to achieve secure transmission. In order to keep the source messages from been eavesdropped, the secrecy capacity should be much smaller than Shannon capacity of the scenarios without eavesdroppers, which introduces multiple antennas into secrecy communications [4–7]. Furthermore, the cooperation for secrecy and cooperative jamming have been studied [8–10]. Generally, it is assumed that the CSI is available at each transmitter, and the secrecy capacity is influenced by different CSI assumptions [11].

In this paper, we focus on secrecy capacity over Rayleigh fading channel with eavesdroppers. Firstly, we propose a simple secure transmission strategy, where the source only communicates with available relay nodes. Here, we only apply the best relay to help the source transmit information to the destination. The outage performance is used to evaluate the system. However, in practical scenarios, the variations of the wireless channel make the obtained CSI becoming outdated. In this case, we come to study the effect of outdated CSI on the system.

© ICST Institute for Computer Sciences, Social Informatics and Telecommunications Engineering 2017
J. Hu et al. (Eds.): SmartGIFT 2016, LNICST 175, pp. 87–97, 2017.
DOI: 10.1007/978-3-319-47729-9_10

The rest of this paper is organized as follows. Section 2 characterizes the system model of the problem. Section 3 analyses the secrecy capacity of wireless channels with eavesdroppers. Section 4 extends the system into outdated CSI scenario. Section 5 provides some numerical results and conclusions are given in Sect. 6.

2   System Model

Let us consider a secrecy communication scenario with two source nodes (one as the source and one as the destination), one eavesdropper and N relays; and all nodes are equipped with a single antenna. We assume that there is no direct link between source-destination, source-eavesdropper, and destination-eavesdropper, respectively. The source transmits information to the destination with the help of N relaying nodes.

The whole information exchanging process can be divided into two phases as shown in Fig. 1. At the first phase, the source broadcast a symbol to N relay nodes under the assumption above. There are two methods to apply achievable relays [12–14]. One is to use all available relays; the other is to use a single relay, which yields the best performance. In this paper, we only focus on the appliance of the best relay to help information transmitting.

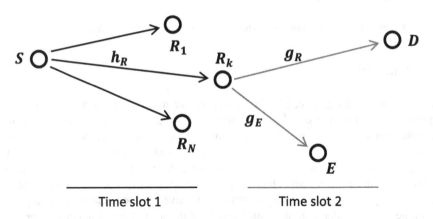

Fig. 1. A diagram for the information transmission strategy with one eavesdropper

Let us choose R_k as the best relay according to the selection criterion in [14], and define h_R, g_R and g_E as the i.i.d. Raleigh fading coefficients for source-relay, relay-destination, and relay-eavesdropper channel, respectively. The received signals at the relay is $y_R = \sqrt{P}h_R s + w_R$, where P is the source transmission power, s is the transmitted signal from the source with unit power, w_R is the AWGN with power P_w.

Based on amplify-and-forward protocol, at the second phase, the relay R_k broadcasts the compressed mixture signal $\frac{\sqrt{P}h_R s + w_R}{\sqrt{P|h_R|^2 + P_w}} \sqrt{P}$ to the destination and the eavesdropper. Hence, the received signals at the destination and eavesdropper can be written as,

$$y_D = \frac{\sqrt{P}h_R s + w_R}{\sqrt{P|h_R|^2 + P_w}} \sqrt{P} \cdot g_R + w_D \tag{1}$$

$$y_E = \frac{\sqrt{P}h_R s + w_R}{\sqrt{P|h_R|^2 + P_w}} \sqrt{P} \cdot g_E + w_E \tag{2}$$

where w_D and w_E denote the additive noise at the destination and eavesdropper with the power P_D and P_E.

3 Secrecy Capacity with Eavesdroppers

In this section, we characterize the secrecy capacity of the system in term of outage probability, $P_{out}(R) = P(C_S < R)$. The secure communication will be ensured only if the secrecy rate C_S is larger than the targeted secrecy data rate R, otherwise the confidential transmission will not be guaranteed.

The capacity of the main channel is,

$$C_M = \frac{1}{2} \log \left(1 + \frac{\rho_R \rho_M |h_R|^2 |g_R|^2}{\rho_M |g_R|^2 + \rho_R |h_R|^2 + 1} \right) \tag{3}$$

And the capacity of the eavesdropper channel is,

$$C_E = \frac{1}{2} \log \left(1 + \frac{\rho_R \rho_E |h_R|^2 |g_E|^2}{\rho_E |g_E|^2 + \rho_R |h_R|^2 + 1} \right) \tag{4}$$

In this paper, we focus on the general case with symmetrical setup, $\rho_M = \rho_E = \rho_R = \rho$.

We can obtain the secrecy capacity as,

$$C_S = C_M - C_E = \frac{1}{2} \log \left(1 + \frac{\rho^2 |h_R|^2 |g_R|^2}{\rho |g_R|^2 + \rho |h_R|^2 + 1} \right) - \frac{1}{2} \log \left(1 + \frac{\rho^2 |h_R|^2 |g_E|^2}{\rho |g_E|^2 + \rho |h_R|^2 + 1} \right) \tag{5}$$

Hence, the outage probability for the targeted secrecy data rate R is written as,

$$P(C_S < R) = P \left(\frac{1 + \frac{\rho^2 |h_R|^2 |g_R|^2}{\rho |g_R|^2 + \rho |h_R|^2 + 1}}{1 + \frac{\rho^2 |h_R|^2 |g_E|^2}{\rho |g_E|^2 + \rho |h_R|^2 + 1}} < 2^{2R} \right) = P \left(\frac{\frac{1}{\rho} + \frac{|h_R|^2 |g_R|^2}{|g_R|^2 + |h_R|^2 + 1/\rho}}{\frac{1}{\rho} + \frac{|h_R|^2 |g_E|^2}{|g_E|^2 + |h_R|^2 + 1/\rho}} < 2^{2R} \right) \tag{6}$$

At high SNR region and for a fixed target rate, the expression of the outage probability can be derived as,

$$P(C_S < R) = P\left(\frac{|g_R|^2}{|g_R|^2 + |h_R|^2} \bigg/ \frac{|g_E|^2}{|g_E|^2 + |h_R|^2} < 2^{2R}\right)$$

$$= P\left(\frac{|h_R|^2|g_R|^2 + |g_R|^2|g_E|^2}{|h_R|^2|g_E|^2 + |g_R|^2|g_E|^2} < 2^{2R}\right) \tag{7}$$

Let us set $A = |h_R|^2$, $x = |g_R|^2$ and $y = |g_E|^2$, and they are exponentially distributed with the parameter λ, λ_x and λ_y respectively. Define $Z = \frac{Ax+xy}{Ay+xy}$, the CDF of Z can be shown as,

$$P(Z < z) = P\left(\frac{Ax+xy}{Ay+xy} < z\right) = \iiint\limits_{\frac{Ax+xy}{Ay+xy} < z} f_A(A) \cdot f_x(x) \cdot f_y(y) dx dy dA$$

$$= \iiint\limits_{\frac{Ax+xy}{Ay+xy} < z} \lambda e^{-\lambda A} \cdot \lambda_x e^{-\lambda_x x} \cdot \lambda_y e^{-\lambda_y y} dx dy dA$$

$$= \int_0^\infty \lambda e^{-\lambda A}\left[\int_{\frac{A}{z-1}}^\infty \lambda_y e^{-\lambda_y y}\int_0^\infty \lambda_x e^{-\lambda_x x} dx dy + \int_0^{\frac{A}{z-1}} \lambda_y e^{-\lambda_y y}\int_0^{\frac{Ayz}{A-y(z-1)}} \lambda_x e^{-\lambda_x x} dx dy\right] dA \tag{8}$$

The part in the square bracket of (8) can be easily simplified as,

$$\int_{\frac{A}{z-1}}^\infty \lambda_y e^{-\lambda_y y}\int_0^\infty \lambda_x e^{-\lambda_x x} dx dy + \int_0^{\frac{A}{z-1}} \lambda_y e^{-\lambda_y y}\int_0^{\frac{Ayz}{A-y(z-1)}} \lambda_x e^{-\lambda_x x} dx dy$$

$$= \int_{\frac{A}{z-1}}^\infty \lambda_y e^{-\lambda_y y} dy + \int_0^{\frac{A}{z-1}} \lambda_y e^{-\lambda_y y}\left[1 - e^{-\frac{\lambda_x Ayz}{A-y(z-1)}}\right] dy \tag{9}$$

$$= \int_{\frac{A}{z-1}}^\infty \lambda_y e^{-\lambda_y y} dy + \int_0^{\frac{A}{z-1}} \lambda_y e^{-\lambda_y y} dy - \int_0^{\frac{A}{z-1}} \lambda_y e^{-\lambda_y y} \cdot e^{-\frac{\lambda_x Ayz}{A-y(z-1)}} dy$$

$$= 1 - \int_0^{\frac{A}{z-1}} \lambda_y e^{-\lambda_y y} \cdot e^{-\frac{\lambda_x Ayz}{A-y(z-1)}} dy$$

For the second part of (9), we can use Gauss-Chebyshev quadrature below to calculate [15].

$$\int_a^b f(y) dy \doteq \frac{\pi(b-a)}{2n}\sum_{i=1}^n f\left(\frac{(1+x_i)(b-a)}{2} + a\right)\left(1 - x_i^2\right)^{1/2} \tag{10}$$

where $x_i = \cos\left(\frac{2i-1}{2n}\pi\right)$, $(i = 1, \ldots, n)$ is the quadrature nodes.

Applying the Eq. (10), we can have

$$\int_0^{\frac{A}{z-1}} \lambda_y e^{-\lambda_y y} \cdot e^{-\frac{\lambda_x Ayz}{A-y(z-1)}} dy = \int_0^{\frac{A}{z-1}} f(y) dy \doteq \frac{\pi \frac{A}{z-1}}{2n} \sum_{i=1}^n f\left(\frac{(1+x_i)\frac{A}{z-1}}{2}\right)(1-x_i^2)^{1/2}$$

$$= \frac{A\pi}{2n(z-1)} \sum_{i=1}^n \lambda_y e^{-\lambda_y \frac{A(1+x_i)}{2(z-1)}} \cdot e^{-\frac{\lambda_x Az(1+x_i)}{(1-x_i)(z-1)}} (1-x_i^2)^{1/2}$$

$$= \frac{\lambda_y A\pi}{2nr} \sum_{i=1}^n e^{-\lambda_y \frac{A(1+x_i)}{2r}} \cdot e^{-\frac{\lambda_x A(1+r)(1+x_i)}{r(1-x_i)}} (1-x_i^2)^{1/2}$$

$$= \frac{\lambda_y A\pi}{2nr} \sum_{i=1}^n (1-x_i^2)^{1/2} \cdot e^{-A\frac{\lambda_y\left(1-x_i^2\right)+2\lambda_x(1+r)(1+x_i)}{2r(1-x_i)}}$$

(11)

where $r = z - 1$.

Combining (8) and (11), the CDF of Z becomes,

$$P(Z < z) = P\left(\frac{Ax+xy}{Ay+xy} < z\right)$$

$$= \int_0^\infty \lambda e^{-\lambda A}\left[1 - \frac{\lambda_y A\pi}{2nr} \sum_{i=1}^n (1-x_i^2)^{1/2} \cdot e^{-A\frac{\lambda_y\left(1-x_i^2\right)+2\lambda_x(1+r)(1+x_i)}{2r(1-x_i)}}\right] dA$$

(12)

$$= 1 - \frac{\lambda\lambda_y \pi}{2nr} \sum_{i=1}^n (1-x_i^2)^{1/2} \int_0^\infty A \cdot e^{-A\frac{\lambda_y\left(1-x_i^2\right)+2\lambda_x(1+r)(1+x_i)+2\lambda r(1-x_i)}{2r(1-x_i)}} dA$$

From Eq. (3.351.3) in [16], and after some algebraic manipulation, we can obtain the outage probability for the targeted secrecy data rate R as,

$$P(C_S < R)$$

$$= 1 - \frac{2\lambda\lambda_y \pi r}{n} \sum_{i=1}^n \frac{(1-x_i^2)^{1/2}(1-x_i)^2}{[2\lambda r(1-x_i)+2\lambda_x(1+r)(1+x_i)+\lambda_y(1-x_i^2)]^2}$$

(13)

where $r = 2^{2R} - 1$.

4 Outdated CSI Scenario

Most existing network systems assume that each transceiver has the perfect CSI. However, in practical use, the actual channel may differ from the outdated one, which makes the information transmission be processed on the outdated CSI. In this case, the overall performance of the whole system will be influenced. In this paper, we assume that only the source-relay channel becomes outdated.

Let us define the correlation coefficient between $\hat{h}_R$ and h_R on the link of the best relay as φ. Adopting Jakes' scattering model, we have the correlation coefficient

$\varphi = J_0(2\pi f_d T_s)$, where $J_0(\cdot)$ denotes the zero-order Bessel function of the first kind, f_d stands for the Doppler frequency, and T_s is the delay in time units [17].

Applying Bayes' theorem, the pdf of h_R is conditioned on $\hat{h}_R$,

$$f_{h_R|\hat{h}_R}\left(h_R|\hat{h}_R\right) = \frac{f_{h_R,\hat{h}_R}\left(h_R,\hat{h}_R\right)}{f_{\hat{h}_R}\left(\hat{h}_R\right)} = \frac{1}{\pi^M \det(G_R)} \cdot e^{-(h_R-\varphi h_R)^H G_R^{-1}\left(h_R-\varphi\hat{h}_R\right)} \tag{14}$$

where $G_R = (1 - \varphi^2)I_M$ is the covariance matrix modelling the degree of CSI uncertainty, and when $h_R = \hat{h}_R$, $G_R = 0$.

Recalling that $A = |h_R|^2$, we can have $\hat{A} = |\hat{h}_R|^2$, and A is conditioned on $\hat{A}$ following a non-central chi-square distribution with two degrees of freedom. We can obtain the conditioned PDF as [18],

$$f_{A|\hat{A}}\left(A|\hat{A}\right) = \frac{\lambda}{1 - \varphi^2} \cdot e^{\frac{-\lambda(A+\varphi^2\hat{A})}{1-\varphi^2}} \cdot I_0\left(\frac{2\lambda\varphi\sqrt{A\hat{A}}}{1 - \varphi^2}\right) \tag{15}$$

where $I_0(\cdot)$ denotes the zero-order modified Bessel function of the first kind.

The PDF of A can be obtained from the marginal distribution density function and conditional distribution as,

$$f_A(A) = \int_0^\infty f_A(A,\hat{A})dA = \int_0^\infty f_{A|\hat{A}}\left(A|\hat{A}\right)dA \cdot f_{\hat{A}}(\hat{A})d\hat{A} \tag{16}$$

Recalling the Eqs. (8) and (12), we can have the actual CDF of Z based on the outdated $\hat{A}$ as,

$$P(Z<z) = \int_0^\infty \iiint_{\frac{\hat{A}x+xy}{\hat{A}y+xy}<z} f_{A|\hat{A}}\left(A|\hat{A}\right) \cdot f_{\hat{A}}(\hat{A}) \cdot f_x(x) \cdot f_y(y)dxdyd\hat{A}dA$$

$$= \int_0^\infty \frac{\lambda^2}{1-\varphi^2} \cdot e^{\frac{-\lambda A}{1-\varphi^2}} \int_0^\infty e^{\frac{-\lambda\hat{A}}{1-\varphi^2}} I_0\left(\frac{2\lambda\varphi\sqrt{A}}{1-\varphi^2}\sqrt{\hat{A}}\right)d\hat{A}dA$$

$$- \int_0^\infty \frac{\lambda^2}{1-\varphi^2} \cdot e^{\frac{-\lambda A}{1-\varphi^2}} \int_0^\infty e^{\frac{-\lambda\hat{A}}{1-\varphi^2}} I_0\left(\frac{2\lambda\varphi\sqrt{A}}{1-\varphi^2}\sqrt{\hat{A}}\right) \cdot \frac{\lambda_y\hat{A}\pi}{2nr} \sum_{i=1}^n (1-x_i^2)^{1/2} \cdot e^{-\hat{A}\frac{\lambda_y(1-x_i^2)+2\lambda_x(1+r)(1+x_i)}{2r(1-x_i)}} d\hat{A}dA$$

$$\tag{17}$$

From the Eq. (8.447.1) in [16], the Bessel function section becomes

$$I_0\left(\frac{2\lambda\varphi\sqrt{A}}{1-\varphi^2}\sqrt{\hat{A}}\right) = \sum_{k=0}^\infty \frac{\lambda^{2k}\varphi^{2k}A^k\hat{A}^k}{(1-\varphi^2)^{2k}(k!)^2}.$$

The first part at the left side of (17) can be easily obtained,

$$\int_0^\infty \frac{\lambda^2}{1-\varphi^2} \cdot e^{\frac{-\lambda A}{1-\varphi^2}} \int_0^\infty e^{\frac{-\lambda \hat{A}}{1-\varphi^2}} \cdot I_0\left(\frac{2\lambda\varphi\sqrt{A}}{1-\varphi^2}\sqrt{\hat{A}}\right) d\hat{A}dA$$

$$= \int_0^\infty \frac{\lambda^2}{1-\varphi^2} \cdot e^{\frac{-\lambda A}{1-\varphi^2}} \int_0^\infty e^{\frac{-\lambda \hat{A}}{1-\varphi^2}} \sum_{k=0}^\infty \frac{\lambda^{2k}\varphi^{2k}A^k\hat{A}^k}{(1-\varphi^2)^{2k}(k!)^2} d\hat{A}dA \qquad (18)$$

$$= \sum_{k=0}^\infty \frac{\lambda^{k+1}\varphi^{2k}}{(1-\varphi^2)^k k!} \int_0^\infty A^k e^{\frac{-\lambda A}{1-\varphi^2}} dA = \sum_{k=0}^\infty \varphi^{2k}(1-\varphi^2) \approx 1$$

For the second part of (17), it is quite complicated to be calculated,

$$\int_0^\infty \frac{\lambda^2}{1-\varphi^2} \cdot e^{\frac{-\lambda A}{1-\varphi^2}} \int_0^\infty e^{\frac{-\lambda \hat{A}}{1-\varphi^2}} I_0\left(\frac{2\lambda\varphi\sqrt{A}}{1-\varphi^2}\sqrt{\hat{A}}\right) \cdot \frac{\lambda_y\hat{A}\pi}{2nr} \sum_{i=1}^n (1-x_i^2)^{1/2} \cdot e^{-\hat{A}\frac{\lambda_y\left(1-x_i^2\right)+2\lambda_x(1+r)(1+x_i)}{2r(1-x_i)}} d\hat{A}dA$$

$$= \frac{\lambda_y\pi}{2nr} \sum_{k=0}^\infty \frac{\lambda^{k+1}\varphi^{2k}(k+1)}{(1-\varphi^2)^k} \sum_{i=1}^n (1-x_i^2)^{1/2}\left(\frac{\lambda}{1-\varphi^2} + \frac{\lambda_y\left(1-x_i^2\right)+2\lambda_x(1+r)(1+x_i)}{2r(1-x_i)}\right)^{-k-2}$$

$$(19)$$

Finally, we can obtain the outage probability for the targeted secrecy data rate R with outdated CSI as,

$$P(C_S < R) = \sum_{k=0}^\infty \varphi^{2k}(1-\varphi^2)$$

$$- \frac{\lambda_y\pi}{2nr} \sum_{k=0}^\infty \frac{\lambda^{k+1}\varphi^{2k}(k+1)}{(1-\varphi^2)^k} \sum_{i=1}^n (1-x_i^2)^{1/2}\left(\frac{\lambda}{1-\varphi^2} + \frac{\lambda_y\left(1-x_i^2\right)+2\lambda_x(1+r)(1+x_i)}{2r(1-x_i)}\right)^{-k-2} \qquad (20)$$

where $r = 2^{2R} - 1$.

5   Numerical Results

In this section, we will provide detailed simulation results to illustrate the performance evaluations in term of outage probability of normal scenario and outdated CSI scenario.

Figure 2 shows the outage probability of the system at different SNR. We use Monte Carlo method to do the simulation, and it is clear that the system can obtain better outage probability performance with the SNR increasing.

In Fig. 3, the curves represent the analytical results comparing with the simulation results at high SNR region. As can be seen, the simulation and analytical approximation are very close, when we choose $n = 10$ Gauss-Chebyshev quadrature nodes. If we use $n = 5$ nodes, the gap between the simulation and analytical results becomes larger at low targeted secrecy data rate.

Figure 4 demonstrates the outage probability of the system at outdated CSI stages. The system performance varies with the correlation coefficient between the actual channel and outdated channel.

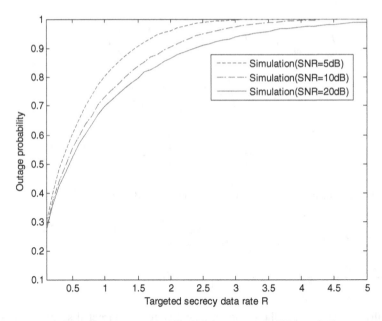

Fig. 2. The outage probability of the system at different SNR

The curves in Fig. 5 also indicate that the number of Gauss-Chebyshev quadrature nodes applied here influence performance of the system.

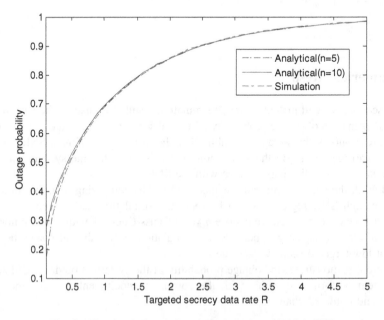

Fig. 3. The simulation and analytical results with high SNR

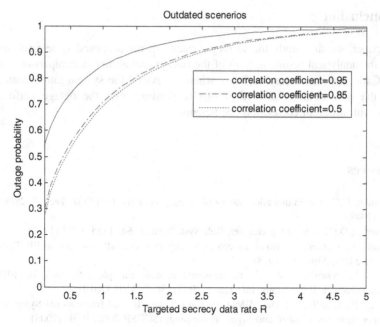

Fig. 4. The outage probability of the system with different correlation coefficients

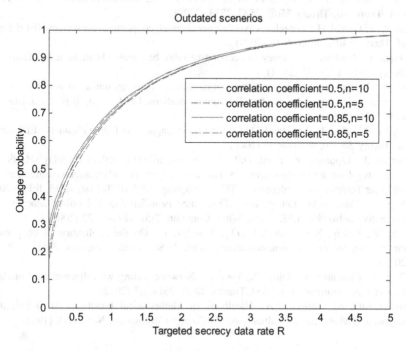

Fig. 5. The outage probability with different correlation coefficients and Gauss-Chebyshev quadrature nodes

6 Conclusion

In this paper, we first study the secrecy capacity of the proposed system scheme and present the analytical results in term of the outage probability in comparison with the Monte Carlo simulation results. Then, when we extend the system into outdated scenarios, the new obtained analytical results illustrate that the outage performance changes with different correlation coefficients.

References

1. Shannon, C.E.: Communication theory of secrecy systems. Bell Syst. Tech. J. **28**(4), 656–715 (1949)
2. Wyner, A.D.: The wire-tap channel. Bell Syst. Tech. J. **54**, 1355–1387 (1975)
3. Csiszár, I., Korner, J.: Broadcast channels with confidential messages. IEEE Trans. Inf. Theory **24**(3), 339–348 (1978)
4. Khisti, A., Wornell, G.W.: Secure transmission with multiple antennas I: the MISOME wiretap channel. IEEE Trans. Inf. Theory **56**(7), 3088–3104 (2010)
5. Oggier, F., Hassibi, B.: The MIMO wiretap channel. In: 3rd International Symposium on Communications, Control and Signal Processing, ISCCSP 2008. IEEE (2008)
6. Khisti, A., Wornell, G.W.: Secure transmission with multiple antennas—Part II: the MIMOME wiretap channel. IEEE Trans. Inf. Theory **56**(11), 5515–5532 (2010)
7. Liu, T., Shamai, S.: A note on the secrecy capacity of the multiple-antenna wiretap channel. IEEE Trans. Inf. Theory **55**(6), 2547–2553 (2009)
8. Lai, L., El Gamal, H.: The relay–eavesdropper channel: cooperation for secrecy. IEEE Trans. Inf. Theory **54**(9), 4005–4019 (2008)
9. Ekrem, E., Ulukus, S.: Secrecy in cooperative relay broadcast channels. IEEE Trans. Inf. Theory **57**(1), 137–155 (2011)
10. He, X., Yener, A.: Two-hop secure communication using an untrusted relay: a case for cooperative jamming. In: Global Telecommunications Conference, IEEE GLOBECOM 2008. IEEE (2008)
11. Gopala, P.K., Lai, L., El Gamal, H.: On the secrecy capacity of fading channels. IEEE Trans. Inf. Theory **54**(10), 4687–4698 (2008)
12. Bletsas, A., Lippnian, A., Reed, D.P.: A simple distributed method for relay selection in cooperative diversity wireless networks, based on reciprocity and channel measurements. In: Vehicular Technology Conference, VTC 2005-Spring. 2005 IEEE 61st, vol. 3. IEEE (2005)
13. Ding, Z., Chin, W.H., Leung, K.K.: Distributed beamforming and power allocation for cooperative networks. IEEE Trans. Wirel. Commun. **7**(5), 1817–1822 (2008)
14. Ding, Z., Leung, K.K., Goeckel, D.L., Towsley, D.: On the application of cooperative transmission to secrecy communications. IEEE J. Sel. Areas Commun. **30**(2), 359–368 (2012)
15. Chen, T., Cumanan, K., Ding, Z., Tian, G.: Network coding with diversity and outdated channel state information. J. Mod. Transp. **20**(4), 261–267 (2012)
16. Abramowitz, M., Stegun, I.A.: Handbook of Mathematical Functions: With Formulas, Graphs, and Mathematical Tables. No. 55. Courier Corporation, New York (1964)

17. Gradshteyn, I.S., Ryzhik, I.M.: Table of Integrals, Series and Products, 6th edn. Academic Press, New York (2000)
18. Vicario, J.L., Anton-Haro, C.: Analytical assessment of multi-user vs. spatial diversity trade-offs with delayed channel state information. IEEE Commun. Lett. **10**(8), 588–590 (2006)
19. Kim, S., Park, S., Hong, D.: Performance analysis of opportunistic relaying scheme with outdated channel information. IEEE Trans. Wirel. Commun. **12**(2), 538–549 (2013)

The Role of Analog Beamwidth in Spectral Efficiency of Millimeter Wave Ad Hoc Networks

Pan Cao[✉] and John S. Thompson

Institute for Digital Communications, The University of Edinburgh,
Edinburgh EH3 9JL, UK
{p.cao,john.thompson}@ed.ac.uk

Abstract. This work considers a millimeter wave (mmWave) ad hoc network, where the analog beam-codebook based beam training is employed to determine the analog beamforming direction for each node. In the analog beam training, a narrow analog beam pattern achieves a high main lobe gain and also reduces the probability of receiving the interference from its main lobe, but it requires high beam training overhead which might in return degrade the spectral efficiency of the network. Thus, there exists a trade-off between the analog beamwidth and the spectral efficiency performance. This motivates us to characterize the spectral efficiency as a function of the analog beamwidth and analyze the role of the beamwidth in the spectral efficiency. Numerical results confirm the effectiveness of the analysis and also provide a suggestion on how to choose the best beamwidth for a given setup.

Keywords: Analog beam codebook · Phase shifter · Beam training · Millimeterwave ad hoc networks · Spectral efficiency

1 Introduction

Consider a millimeterwave (mmWave) ad hoc network, where each node has a linear antenna array with a single radio frequency (RF) and analog-to-digital convector (ADC) chain. This power/cost-efficient hardware architecture captures the sparse property of a mmWave channel [1] and also enables the simple implementation of the analog beamforming design (phase shifter). In order to improve the spectral efficiency, the transmit/receive beamforming should be designed in a closed-loop fashion to adapt to the varying channel state information (CSI). However, the traditional pilot transmission based channel estimation and beamforming design policy in microwave systems is not applicable to the mmWave systems with large scale antenna arrays because the high overhead/time required conflicts with the short coherence time of the mmWave channel. By making use of the sparse property of the mmWave channels [1], analog beam-codebook based beam training is preferred due to its simple implementation. The basic idea is

P. Cao and J. Thompson acknowledge financial support for this research from the UK EPSRC grant number EP/L026147/1.

© ICST Institute for Computer Sciences, Social Informatics and Telecommunications Engineering 2017
J. Hu et al. (Eds.): SmartGIFT 2016, LNICST 175, pp. 98–104, 2017.
DOI: 10.1007/978-3-319-47729-9_11

that each node transmit/receive the signal by using different beam codewords in a codebook, where the beam codebook(s) can be generated offline and stored at the transmitters and the receivers. After an exhaustive "beam sweeping", the best beam pair can be determined based on the received signal strength [2].

Based on the beam training, the authors in [3] provide the coverage and capacity analysis for mmWave cellular networks. Some previous work on mmWave ad hoc networks is largely restricted to indoor scenarios with short range, e.g., [4]. In [5], the authors use a stochastic geometry approach to characterize the one-way and two-way signal-to-interference-plus-noise ratio (SINR) distribution for an outdoor mmWave ad hoc network. In [6], the mmWave ad hoc network performance might be noise-limited and interference-limited, which depends on the nodes density. By employing the analog beam pattern designed by the shaped beam synthesis approach proposed in [7], the main beam lobe gain becomes greater when the beamwidth becomes narrower, and the probability of the interference coming/going through the main beam lobe of a receiver/transmitter becomes smaller, which will enhance the SINR of each link if the beam alignment is perfect. However, the narrower beamwidth leads to more overhead and time consumption during the beam training such that less time is left for the data transmission, which might degrade the spectral efficiency. Therefore, there exists a trade-off between the beamwidth and the spectral efficiency.

However, little previous work provides a analyze the role of the analog beamwidth in the spectral efficiency of the network. This motivates us to find answer for this problem. By employing the beam codebooks generated by [7], we characterize the beam gain as a function of the beamwidth, and then we derive a closed-form expression to approximate the average spectral efficiency with respect to the beamwidth. This novel result is the first work that provides a way to determine the near-optimal beamwidth for a given mmWave ad hoc network.

2   System Model

Consider an ad hoc mmWave network consisting of K pairs of transmitters (Tx) and receivers (Rx). For each pair, a Tx wishes to send data symbols to a Rx using the same time slots and frequency spectrum as the other pairs, and thus this ad hoc network can be modeled as a K-user (pair) mmWave interference channel. The point-to-point communication of the k-th user is denoted by Tx $k \mapsto$ Rx k. To overcome the severe path loss on mmWave bands and meanwhile minimize the hardware complexity, we assume that each node is equipped with an N-antenna linear array associated with a single RF and ADC chain. For the narrow-band and far-field transmission, the array response vector of an N-antenna array at Tx k or Rx k can be defined as

$$a(\theta) = \left[1, e^{j2\pi \frac{r(1)}{\lambda} \cos\theta}, \cdots, e^{j2\pi \frac{r(N-1)}{\lambda} \cos\theta}\right]^T, \tag{1}$$

where the $(N-1) \times 1$ vector r contains the distances from the first antenna (reference element) to the other $N_k - 1$ antennas; and $\theta \in \Theta$ denotes the angle of departure (AoD) of the signal from Tx k or the angle of arrival (AoA) to Rx k where $\Theta = [0°, 180°]$ is defined as the physical angles that cover the entire (one-sided) spatial horizon for a linear array.

For the above setup, analog beamforming is preferred so the phase shifter can control the radiated beam pattern such that the main portion can be concentrated to the desired direction in the spatial domain. Let $f_k \in C^{N \times 1}$ and $g_k \in C^{N \times 1}$ be the two *constant-modulus* and *unit-norm* analog beamforming vectors employed at Tx k and Rx k, respectively. Then, the received signal at Rx k can be expressed as

$$y_k = \sqrt{p_k} g_k^H H_{kk} f_k s_k + \sum_{\ell \neq k} \sqrt{p_\ell} g_k^H H_{k\ell} f_\ell s_\ell + g_k^H z_k \qquad (2)$$

where $H_{kk} \in C^{N \times N}$ and $H_{k\ell} \in C^{N \times N}, \ell \neq k$ denote the channel matrices of the desired link Tx $k \mapsto$ Rx k, and the interfering link Tx $\ell \mapsto$ Rx k, respectively; and s_k with $E\{|s_k|\} = 1$ is the desired transmit symbol from Tx k to Rx k by the transmit power $\sqrt{p_k}$, which is subject to a maximum transmit power constraint of $p_k \leq \overline{p}_k$. The vector $z_k \in C^{N \times 1}$ denotes the thermal noise at Rx k and satisfies the distribution of $\mathcal{CN}(0, \sigma^2 BI)$, where σ^2 denotes the thermal noise floor for 1 Hz bandwidth and B Hz denotes the occupied spectrum bandwidth.

We adopt a widely-used narrowband clustered channel representation to model a mmWave channel, i.e., [1]

$$H_{k\ell} = \underbrace{\mathcal{I}_{LOS} \alpha_{k\ell,LOS} a_k(\phi_{LOS}) a_\ell^H(\psi_{LOS})}_{LOS\ ray} + \underbrace{\sum_{m=1}^{M_{k\ell}} \alpha_{k\ell,m} a_k(\phi_m) a_\ell^H(\psi_m)}_{NLOS\ rays}, \qquad (3)$$

where $\mathcal{I}_{LOS} \in \{0, 1\}$ is used to denote the bloackage of the LOS link or not. Without loss of generality, we assume that there exist $M_{k\ell}$ NLOS rays and $M_{k\ell}$ depends on the scattering environment and is small in general for mmWave system. The parameters $\alpha_{k\ell,LOS}$ and $\{\alpha_{k\ell,m}\}_{m=1}^{M_{k\ell}}$ denote the complex gain of LOS link and NLOS links, respectively. Based on the statistical model for mmWave channels suggested in [1], each complex gain can be expressed as $\alpha_{k\ell,m} = \sqrt{\rho_{k\ell,m}} h_{k\ell,m}$ where $\rho_{k\ell,m}$ and $h_{k\ell,m} \sim \mathcal{CN}(0, 1)$ denotes the large scale fading and small scale fading of the m-th ray. More precisely, the LOS large scale fading can be described as:

$$\rho_{k\ell,LOS} = 10^{-0.1 a_{LOS}} d_{k\ell}^{-2}, \qquad (4)$$

where $a_{LOS} = 20 \log_{10}(f_c) + 32.45$ denotes the parameters of the free-space path loss model where f_c is the carrier frequency measured in units of GHz, e.g., $a_{LOS} = 64.5$ when $f_c = 28\,\text{GHz}$. Similarly, the total NLOS large scale fading can be defined as

$$\rho_{k\ell,NLOS} = \sum_{m=1}^{M_{k\ell}} \rho_{k\ell,m} = 10^{-0.1 a_{NLOS}} d_{k\ell}^{-b_{NLOS}}, \qquad (5)$$

where the parameters a_{NLOS} and b_{NLOS} are shown in [1, Table I].

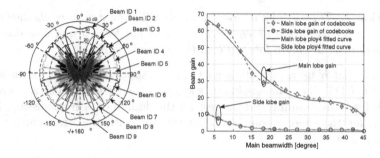

Fig. 1. Left: A 9-beam codebook ($\overline{\theta} = 20°$) for a 64-antenna ULA generated based on [7]; Right: average beam gain vs. beamwidth for a 64-antenna ULA

According to the definition of the array response vector in (1), $\phi_{k\ell,m}$ and $\psi_{k\ell,m}$ refer to the AoA to Rx k and the AoD from Tx k of the m-th ray, respectively. We assume that the LOS ray of Tx $\ell \mapsto$ Rx k is blocked with the probability (w.p.) of $P_{k\ell,NLOS} \in [0,1]$. For example, the probability of blockage of the LOS ray can be defined as [1] $P_{k\ell,NLOS} = 1 - e^{-c_{LOS}d_{k\ell}}$, where c_{LOS} is a coefficient suggested in [1, Table I]. This distance-dependent definition is based on the fact that the probability of the blockage of the LOS ray is increasing with the distance of the link.

3   Beam Codebook Based Beam Training

For the single resolution beam training, the Tx and the Rx need to exhaustively transmit and receive via different beams in the codebook and then to determine the best transmit/receive beam pair based on the received signal strength [2,4]. This allows beam codebook(s) to be generated offline. In this work, the analog beam codebooks are designed based on the shaped beam synthesis approach proposed in our work [7], which could generate a beam pattern with a flat main lobe. For example, Fig. 1-Left shows a 9-beam codebook example with the main beamwidth 20° for a 64-antenna uniform linear array (ULA). Let $\overline{\theta}$, $G(\overline{\theta})$ and $g(\overline{\theta})$ denote the main beamwidth, average main lobe gain and average side lobe gain, respectively. From Fig. 1-Right, $G(\overline{\theta})$ and $g(\overline{\theta})$ are decreasing with the main beamwidth $\overline{\theta}$, and the average main lobe gain and the side lobe gain can be near-perfectly fitted to the following *biquadratic functions* of $\overline{\theta} \in [3°, 45°]$:

$$G(\overline{\theta}) = -0.0001082\overline{\theta}^4 + 0.009862\overline{\theta}^3 - 0.2558\overline{\theta}^2 + 0.07395\overline{\theta} + 68.19 \quad (6)$$
$$g(\overline{\theta}) = 6.261 \times 10^{-6}\overline{\theta}^4 - 0.001098\overline{\theta}^3 + 0.6494\overline{\theta}^2 - 1.585\overline{\theta} + 14.69. \quad (7)$$

In order to measure the influence of the beam training procedure on the spectral efficiency, We introduce a new concept – the *coherent operation time (COT)* for a mmWave channel as follows:

Definition 1. *The* coherent operation time (COT) *of a mmWave channel refers to the time period during which the LOS ray is not blocked and also this LOS ray goes through the main lobes of both the Tx and the Rx.*

We assume that K users have the same COT and support τ symbol transmission. During the COT, the transmission of τ symbols is split to two parts: β symbol intervals for beam training and the rest for the desired data transmission. More precisely, for the single-resolution beam training with the beamwidth $\bar{\theta}$, the required overhead is[1]

$$\beta = (180/\bar{\theta})^2 \qquad (8)$$

symbols transmission during the beam training.

4   Spectral Efficiency Performance Analysis

From (2), the transmission rate of link TX $k \mapsto$ RX k can be expressed as

$$R_k = \left(1 - \frac{\beta}{\tau}\right) B \log_2 \left(1 + SINR_k\right) \quad symbol/second/Hz \qquad (9)$$

where $1 - \frac{\beta}{\tau} < 1$ denotes the portion of each COT for the desired data transmission; and $SINR_k$ denotes the SINR of the k-th user and is expressed as

$$SINR_k = \frac{p_k |g_k^H H_{kk} f_k|^2}{\sum_{\ell \neq k} p_\ell |g_k^H H_{k\ell} f_\ell|^2 + B\sigma^2}. \qquad (10)$$

It is interesting to consider the average SINR over the small scale fading, AoAs and DoDs during the coherence time of the large scale fading. Define $R_s = \sum_{k=1}^{K} E[R_k]$ as the average sum rate over the small scale fading of the K-user ad hoc network, which can be approximately expressed as follows.

Proposition 1. *The average rate R_{sum} can be approximately expressed as*

$$R_s \approx \left(1 - \frac{\bar{\theta}^2}{180^2 \tau}\right) B \sum_{k=1}^{K} \log_2 \left(1 + \frac{p_k E[S_{kk}]}{\sum_{\ell \neq k} p_\ell E[I_{k\ell}] + \sigma^2 B}\right) \qquad (11)$$

where the average desired channel gain and interference channel gain are

$$E[S_{kk}] = (1 - P_{kk,NLOS})G^2(\bar{\theta})\rho_{kk,LOS} + P_{kk,NLOS}\tilde{G}(\bar{\theta})\rho_{k\ell,NLOS} \qquad (12)$$

$$E[I_{k\ell}] = (1 - P_{k\ell,NLOS})\tilde{G}(\bar{\theta})(\rho_{k\ell,LOS} + \rho_{k\ell,NLOS})$$
$$+ P_{k\ell,NLOS}\tilde{G}(\bar{\theta})\rho_{k\ell,NLOS} \qquad (13)$$

---

[1] Different interfering users can operate on different orthogonal frequency band during the beam training in order to avoid receiving the interference [8].

where $\rho_{k\ell,LOS}$ and $\rho_{k\ell,NLOS}$ are defined in (4) and (5), respectively, and

$$\tilde{G}(\bar{\theta}) = G^2(\bar{\theta})\left(\frac{\bar{\theta}}{180^\circ}\right)^2 + G(\bar{\theta})g(\bar{\theta})\frac{\bar{\theta}(180^\circ - \bar{\theta})}{(180^\circ)^2} + g^2(\bar{\theta})\left(\frac{180^\circ - \bar{\theta}}{180^\circ}\right)^2 \quad (14)$$

denotes the effective channel gain averaged on the random AoAs and DoDs of the rays.

Proof. The derivation of (11) is based on the widely-used approximation $E\left[\log_2(1 + \frac{f_1(x)}{f_2(x)})\right] \approx \log_2\left(1 + \frac{E[f_1(x)]}{E[f_2(x)]}\right)$. Therefore, it is left to compute the average desired signal power and interference power in (12) and (13).

If the LOS ray is not blocked, $\alpha_{k\ell,LOS}$ will always experience the largest beam gain after the perfect beam alignment, i.e., $G^2(\bar{\theta})$ where $\bar{\theta}$ denotes the analog beamwidth. Otherwise, some NLOS rays will experience the main lobes of both the Tx and the Rx, and others go through the side lobes. Thus, considering the independency of each NLOS rays, the average antenna gain can be computed over their random AoAs and DoDs, i.e., (14). Then, (12) and (13) are derived.

Remark 1. Plugging (6) and (7) into (14), the average sum rate expression (11) of an ad hoc network becomes a closed-form function of the analog beamwidth $\bar{\theta}$. This novel result enables theoretical analysis of the impact of the beamwidth on spectral efficiency.

Now, the original spectral efficiency maximization problem by optimizing analog beamforming vectors can be equivalently formulated to the problem with respect to $\bar{\theta}$, i.e.,

$$\max_{\{f_k, g_k\}} R_s \quad \Leftrightarrow \quad \max_{\bar{\theta}} R_s, \quad (15)$$

which can be easily solved by a grid line search of $\bar{\theta} \in [3^\circ, 45^\circ]$ to determine the optimal beamwidth.

5   Numerical Results

The spectral efficiency of an ad hoc network is illustrated based on 1000 runs of Monte Carlo simulations, where K Txs and K Rxs are randomly distributed in an area $\mathcal{A} = [-50, 50]\,\text{m} \times [-50, 50]\,\text{m}$. Each node has a 64-antenna ULA with the same transmit power $p_k = 23$ dBm, $\forall k$, and the thermal noise floor is set to be $\sigma^2 = -174$ dBm/Hz. The carrier frequency is 28 GHz. The channel fading coefficients in (4) are set as $a_{LOS} = 64.5$ dB and $c_{LOS} = 1/67.1\,\text{m}$ and $a_{NLOS} = 72$ and $b_{NLOS} = 2.92$.

Figure 2 provides an evaluation of the sum spectral efficiency with respect to the user density in $\mathcal{A}$. The spectrum bandwidth is set as $B = 250\,\text{MHz}$. We observe that the sum spectrum efficiency also achieves the maximum value when $\bar{\theta} = 12^\circ$ and its trend does not vary with the user density. The average sum spectral efficiency is increasing with the user density but the spectral efficiency per user is decreasing with the user density, since the Rx will receives more interference when the user density becomes higher. The theoretical analysis of the role of the user density will be considered in our future work.

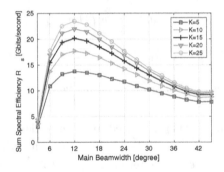

Fig. 2. Sum spectral efficiency vs. beamwidth with different user density

6   Conclusions

In this paper, we study the spectral efficiency of a mmWave ad hoc network. The effect of the analog beamwidth on the beam training and also the spectral efficiency performance is analyzed. In particular, we characterize the average spectral efficiency as a closed-form function of the beamwidth, which provides a theoretical view on the role of the beamwidth in the spectral efficiency performance of a mmWave ad hoc system. The best choice of the beamwidth for a given setup can be efficiently determined based on numerical results.

References

1. Akdeniz, M.R., Liu, Y., Samimi, M.K., Sun, S., Rangan, S., Rappaport, T.S., Erkip, E.: Millimeter wave channel modeling and cellular capacity evaluation. IEEE J. Sel. Areas Commun. **32**(6), 1164–1179 (2014)
2. Dai, F., Jie, W.: Efficient broadcasting in ad hoc wireless networks using directional antennas. IEEE Trans. Parallel Distrib. Syst. **17**(4), 335–347 (2006)
3. Bai, T., Alkhateeb, A., Heath, R.W.: Coverage and capacity of millimeter-wave cellular networks. IEEE Commun. Mag. **52**(9), 70–77 (2014)
4. Baykas, T., Sum, C.S., Lan, Z., Wang, J., Rahman, M.A., Harada, H., Kato, S.: IEEE 802.15.3c: the first IEEE wireless standard for data rates over 1 Gb/s. IEEE Commun. Mag. **49**(7), 114–121 (2011)
5. Thornburg, A., Bai, T., Heath Jr., R.W.: Performance analysis of mmwave ad hoc networks. IEEE Trans. Signal Process. **64**(15), 4065–4079 (2016)
6. Shokri-Ghadikolaei, H., Fischione, C.: Millimeter wave ad hoc networks: noise-limited or interference-limited? In: 2015 IEEE Globecom Workshops (GC Wkshps), pp. 1–7, December 2015
7. Cao, P., Thompson, J., Haas, H.: Constant-modulus shaped Beam synthesis via convex relaxation. IEEE Antennas Wirel. Propag. Lett. **PP**(99), 1–4 (2016)
8. Han, S., Chih-Lin, I., Xu, Z., Wang, S.: Reference signals design for hybrid analog and digital beamforming. IEEE Commun. Lett. **18**(7), 1191–1193 (2014)

MASTERING Workshop

An ANN-Based Energy Forecasting Framework for the District Level Smart Grids

Baris Yuce[✉], Monjur Mourshed, and Yacine Rezgui

School of Engineering, BRE Centre for Sustainable Engineering,
Cardiff University, Queen's Building, The Parade, Cardiff CF24 3AA, UK
{Yuceb,MourshedM,RezguiY}@cardiff.ac.uk

Abstract. This study presents an Artificial Neural Network (ANN) based district level smart grid forecasting framework for predicting both aggregated and disaggregated electricity demand from consumers, developed for use in a low-voltage smart electricity grid. To generate the proposed framework, several experimental study have been conducted to determine the best performing ANN. The framework was tested on a micro grid, comprising six buildings with different occupancy patterns. Results suggested an average percentage accuracy of about 96%, illustrating the suitability of the framework for implementation.

Keywords: ANN · District energy management · Grid electricity · Smart city

1 Introduction

There is an urgent need for transforming the electricity grid management to keep up with the rapid advances in generation technologies, electric power systems and increased integration of distribution energy resources in the low- and medium-voltage (LV/MV) grids [1, 2]. At the heart of this transformation is the integration of ICT and energy infrastructures for providing a user-oriented service in an increasingly decentralized grid where (a) electricity from renewable sources is shared in the LV grid and (b) flexibility is managed, to maximize the utility of the service [2]. Conventional static grid, where energy flows from generation units to consumer [3] is being replaced by bi-directional flows of energy and information, resulting in end-users being metamorphosed into '*prosumers*'; i.e. both a consumer and producer of electricity using photovoltaic (PV), wind, combined heat and power (CHP) technologies [2, 4], thereby increasing the complexity in managing the grid. To deal with this complexity, the smart grid management concept has been stand out as an adaptive solution to deal with all these complexities. Further, the grid management has also been enhanced to district level which increases the complexity of problem higher levels. Patti et al., [5] proposed a district information modelling and energy management system to control the district energy usage according to the user behavior such level of the energy management is

B. Yuce—The work has been funded by the European Commission in the context of the MAS2TERING project (the grant number 619682).

© ICST Institute for Computer Sciences, Social Informatics and Telecommunications Engineering 2017
J. Hu et al. (Eds.): SmartGIFT 2016, LNICST 175, pp. 107–117, 2017.
DOI: 10.1007/978-3-319-47729-9_12

highly complex, and has an unpredictable energy consumption pattern. Therefore, it requires intelligent solution to overcome with this types of difficulties. As intelligent system has been successfully utilized in the built environment to control and predict energy consumption such as artificial neural network [6, 7] genetic algorithm [8], rule based systems [9], and ontology based systems [10].

In this paper, an ANN based forecasting model is presented to predict the energy consumption of a smart grid. The forecasting model is aimed to determine next 15 min energy consumption according to current energy load in the grid. The paper consists of the following sections; the background, ANN based forecasting model, experiments and conclusion sections in Sects. 2, 3, 4 and 5 respectively.

2 Background

District level grid energy management became one of the most popular topic in the area of the smart grid energy management and built environment. In district level energy management, the idea is to satisfy energy requirement for each building connected to the related district [11]. However, it is not easy task to carry out this process without having a robust forecasting model. As the forecasting models are able to provide a vision to future energy consumption of each individual buildings in the related district, which allow to control and manage the entire district well and also reduce the energy consumption according to the needs while maintaining the thermal comfort levels. The generalized hierarchical smart grid energy management can be illustrated as in Fig. 1.

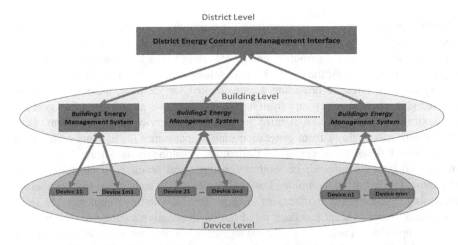

Fig. 1. The hierarchical energy management levels in the smart grid.

As highlighted in Fig. 1, the district level energy control and management process consists of three levels, the first and bottom level is the device based control and management process which provides the communication and energy management of individual devices.

In the second level, the building level energy management and control process appears, and the Building Information and Management (BIM) systems play the key role. As this system provides a holistic management approach to control, manage and visualize the building energy management. The traditional BIM systems have less flexibility in this sense, which contain simple rules to control entire building devices. However, the simplicity may not provide and efficient level of control in the entire buildings. Therefore, several intelligent based solutions have been proposed in the literature to generate much efficient and flexible solutions. Kalogirou [12] proposed an Artificial Neural Network (ANN) - based solution to predict the energy consumption and generation in the renewable sources in building level to control energy consumption in the building well. Moreover, Ferreira et al. [13] proposed another ANN-Genetic Algorithm (GA) - based solution to control the heating ventilation and air conditioning (HVAC) systems while maintaining the thermal comfort level. Furthermore, Gonzalez and Zamarreno [14] also proposed another a short term load prediction method using ANN in the building environment.

As several studies have been utilized using the intelligent system technologies to control energy consumptions in the building level. However, the new technological developments provide to extend this level to the third level which is the district level. The district level of energy control and management is highly dependent on the controlling of energy requirement of each individual. In this stage, the energy control and management process expands to higher level complicated problems. Therefore, the usage of intelligent systems in this scale is becoming a needs to overcome several issues such as robustness, flexibility, adaptiveness and autonomous decision making process. These abilities provide to control entire grid well and optimize the energy consumption very well [15]. Valerio et al. [15] proposed a district level energy optimizer to control the electricity usage in the grid, and to optimize the prosumer's electricity usage according to the forecasting results. Thus, the forecasting module becomes very critical in the district energy management level due to the simplicity compare to the physical model. A well-defined forecasting model can be deployed into optimizer module to control and optimize the entire electricity grid system well. Therefore, it is highly important to get highly accurate forecasting model to predict each individual prosumers' surplus and demanded electricity on site. Therefore, it will provide a prior knowledge organize entire grid based on that.

In this paper, it is presented an ANN based forecasting system for the MAS^{2-}TERING project to determine the individual buildings energy needs which provides the aggregated forecasting results to determine the energy needs of the entire smart grid. This information is then will be utilized in an optimizer agent to optimize the district level energy solution.

3 ANN-Based Forecasting Module

Artificial Neural Network (ANN) is one of the most popular forecasting methods in the area of robotics, machine vision, energy management and control, medicine and manufacturing and social science [9]. This is highly related to the capability of the ANN on the complex and intrinsic problems. As a well-trained ANN is able to predict the outputs of complex problems without using any analytical relationships among the inputs and outputs [7, 9].

In this paper, a forecasting service has been developed using ANN for an EC funded project called MAS^{2}TERING. The forecasting service is utilized Irish Smart Grid data [16] to demonstrate the capability of the proposed model on the smart grid in district level. In the proposed model, six different types' buildings according to the occupancy types are considered to illustrate the strength of the forecasting algorithm on different energy consumers. As the occupancy types is one of the key variables to analyze the energy consumption in the buildings [17]. Thus, it has been selected six different buildings and three of them have different occupancy types such as, three buildings have no child live in the building (numbered as building 1, 4 and 6), one building has a family (two adults) with one child (child age below the age of 15) - (building numbered as 5), the second types of building consists of a family with 2 children who are age of below 15 - (building numbered as 3) and the last type of the building has a family with 3 children (ages are below 15) - (building numbered as 2). To predict the each building energy consumption, individual ANN models are generated for each buildings. Each ANN model consists of current energy consumption with time information and external weather conditions. The outputs of each ANN is the next 15 min energy consumption, the proposed ANN topology is given Fig. 2.

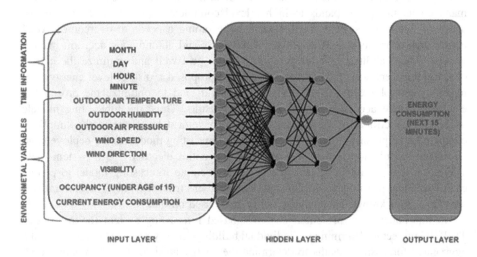

Fig. 2. The proposed ANN topology for each building.

As highlighted in Fig. 2, each ANN has external weather conditions to utilize as input and increase correlation between inputs and output, these weather conditions are selected as outdoor weather temperature, outdoor humidity, outdoor air pressure, wind speed, wind direction and visibility. Moreover, one social variable is also included as input of ANN which is the existence of the children under age 15 in the selected building. This variables has been determined through a sensitivity analysis process. The variables which are related the environmental conditions has a direct impact on the energy consumption such as in highly windy and cold conditions, the energy consumptions tend to go up. Moreover, there is a direct correlation between consumption and the occupants under age of below 15 which generates an expected patterns in the electricity consumption profile. Thus, this property transform the electricity consumption into highly irregular and unpredictable conditions. Finally, time information is also included as month, date, hour and minute in the inputs part of the ANN aligned with the current energy consumption of the selected building. The proposed ANN models for the selected six pilot are utilized to determine the entire districts required electricity demand by aggregation of individuals, shown as in Fig. 3. Each building's ANN performs individually to predict electricity demand, which are then aggregated to produce the district electricity demand. The experiments and results for the selected pilot district are presented in the following sections.

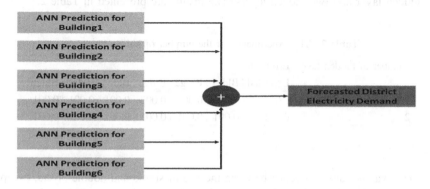

Fig. 3. The proposed forecasted district electricity demand.

4 Experiments

To determine the best performed ANN for each individual building, several experiments were conducted using the Irish Smart Grid data which contains 1.5 years of energy consumption every 30 min. The experiments were carried out to determine best performed ANN topology which requires to determine the best performed the learning function type, the number of hidden layer, the number of process element in hidden layers, and transfer function types in both hidden layers and output layer. The maximum iteration number was set to 5000 epochs. To determine the best performed learning function, the following topology was utilized; two hidden layers with 20 number process elements in each hidden layer and logarithmic sigmoid function in both hidden layers and outputs layers. The following learning algorithms were utilized for experiments on the MATLAB platform; Traingd, Trainbfg, Traincgb, Trainlm, Trainscg [7]. The results are presented in Table 1.

The best performing learning function is found with Levenberg–Marquardt based backpropagation algorithm. Hereafter, this algorithm was utilized as the learning function for the ANN to perform the remaining experiments. The second stage of the

Table 1. The experiments for the learning function types.

Learning function	Error rate						
	Expected	Bldg1	Bldg2	Bldg3	Bldg4	Bldg5	Bldg6
Traingd	0.001	0.031	0.171	0.149	0.094	0.114	0.062
Trainbfg	0.001	0.004	0.201	0.121	0.094	0.329	0.088
Traincgb	0.001	0.006	0.219	0.184	0.051	0.091	0.047
Trainlm	0.001	0.001	0.001	0.001	0.001	0.001	0.001
Trainscg	0.001	0.003	0.167	0.004	0.005	0.002	0.004

*Bldg1, 2, 3, 4, 5, 6: Building number 1, 2, 3, 4, 5, 6.

experiments was to determine the number of process elements in the hidden layer. To carry out this experiments, two types of hidden layer was tested for each building as one hidden layer and two hidden layers. The results are presented in Table 2.

Table 2. The experiments for the number of hidden layer.

Number of hidden layer	Error rate						
	Expected	Bldg1	Bldg2	Bldg3	Bldg4	Bldg5	Bldg6
1	0.001	0.001	0.002	0.002	0.001	0.001	0.001
2	0.001	0.001	0.001	0.001	0.001	0.001	0.001

There was not any difference between the use of single and double layers, except for buildings 2 and 3, where the use of single layer did not meet the criteria for expected error rate. Therefore, two hidden layers were used for subsequent experiments. In third stage of the experiment, the process is carried out to determine the transfer function types in both hidden layer and output layers. It has been utilized logarithmic sigmoid function (LS), hyperbolic tangent sigmoid (TS) and linear (PL) in both hidden layers and output layers. The results are illustrated as in Table 3.

Table 3. The experiments for the transfer function types.

Transfer functions [H1H2 O]	Error rate						
	Expected	Bldg1	Bldg2	Bldg3	Bldg4	Bldg5	Bldg6
LS LS LS	0.001	0.001	0.001	0.001	0.001	0.001	0.001
TS TS TS	0.001	0.007	0.089	0.008	0.013	0.007	0.004
PL PL PL	0.001	0.101	0.190	0.150	0.153	0.145	0.127

*H1: Hidden Layer 1, H2: Hidden Layer 2, O: Output Layer.

According to Table 3, the usage of logarithmic sigmoid function provides the best performed topology. The following experiments will be carried out using this transfer function type in both hidden layers and output layer. The last experiments to determine the best performed topology is to determine the number of process element in the hidden layers. The results are presented in Table 4.

Table 4. The experiments for the number of process elements in the hidden layers.

Transfer functions [H1H2]	Error rate						
	Expected	Bldg1	Bldg2	Bldg3	Bldg4	Bldg5	Bldg6
5 5	0.001	0.073	0.110	0.008	0.103	0.009	0.009
10 10	0.001	0.002	0.048	0.002	0.002	0.008	0.003
20 20	0.001	0.001	0.001	0.001	0.001	0.001	0.001
30 30	0.01	0.006	0.023	0.009	0.009	0.007	0.006

According to the Table 4, the best performed network was found as using 20 process elements in both hidden layers. Based on the entire topology determination process, the performed network utilized to conduct the forecasting process for the district level. According the experimental results, the average forecasting accuracy for each buildings was found as 95.94%, 84.40%, 94.65%, 94.47%, 95.79% and 95.84% by eliminating zero energy consumed days. The results are presented in Figs. 4, 5, 6, 7, 8 and 9.

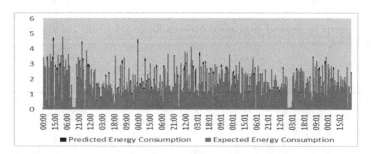

Fig. 4. The electricity demand comparison between forecasted results and expected results for building 1.

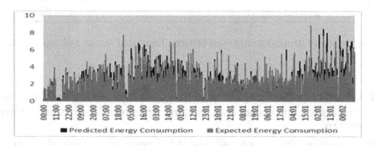

Fig. 5. The electricity demand comparison between forecasted results and expected results for building 2.

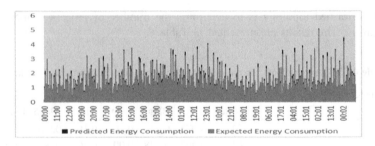

Fig. 6. The electricity demand comparison between forecasted results and expected results for building 3.

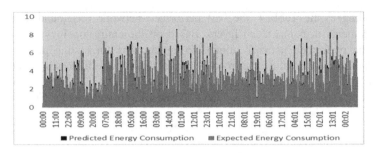

Fig. 7. The electricity demand comparison between forecasted results and expected results for building 4.

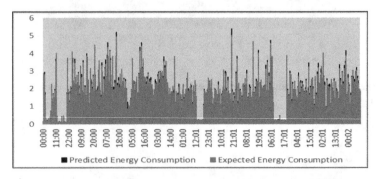

Fig. 8. The electricity demand comparison between forecasted results and expected results for building 5.

Further the average percentage error of the aggregated demand was found as 3.9%. The entire process was integrated under the forecasting interface by Cardiff University, shown in Fig. 10.

As highlighted in Fig. 3, each buildings ANN performs individually to predict the electricity demand for each individual building, then aggregation of entire buildings

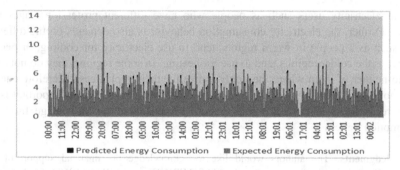

Fig. 9. The electricity demand comparison between forecasted results and expected results for building 6.

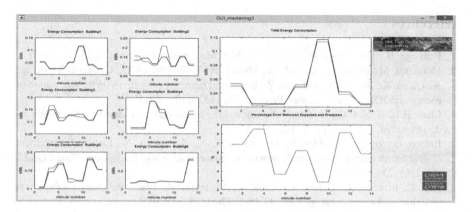

Fig. 10. The forecasting interface.

demand provides the desired demand for the district electricity demand. In the next stage, the experiments and results will be presented for the selected pilot district according to the proposed model.

5 Conclusion

In this study, an ANN based district grid energy demand forecasting model is proposed. To test the performance, the proposed framework, six buildings with different energy characteristic are selected from Irish Smart Grid data [16]. To determine the best performed ANN for each individual building, several different types of ANN topology were tested and presented in the experiments section. According to the results, the district level energy consumption was determined about 4% average percentage error. This figure is also highly related to electricity consumption pattern. As the regular user has a repetitive pattern which allow the intelligent system such as ANN to learn patter quicker with higher prediction rate. However this prediction profile may change with different electricity users such as people who have different professions, ages and income. Further, the electricity consumption behavior is also depends on the different states such as a people in warm regions tend to use electricity for cooling, or people who live is the colder regions tend to use for heating. In some regions, they do not even need to neither of them where they only utilize for their appliances. Therefore energy consumption pattern may derivate from each other, however the main important issue with forecasting is to have a regular pattern to generalize the knowledge from the consumption pattern.

Acknowledgments. The authors would like to acknowledge the financial support of the European Commission in the context of the MAS^{2}TERING project (Ref: 619682) funded under the ICT-2013.6.1 - Smart Energy Grids program.

References

1. Farhangi, H.: The path of the smart grid. IEEE Mag. Power Energy **8**(1), 18–28 (2010)
2. Mourshed, M., Robert, S., Ranalli, A., Messervey, T., Reforgiato, D., Contreau, R., Becue, A., Quinn, K., Rezgui, Y., Lennard, Z.: Smart grid futures: perspectives on the integration of energy and ICT services. Energy Procedia **75**, 1132–1137 (2015)
3. Chao, H.L., Tsai, C.C., Hsiung, P.A., Chou, I.H.: Smart grid as a service: a discussion on design issues. Sci. World J. **2014**, 1–11 (2014)
4. Brusco, G., Burgio, A., Menniti, D., Pinnarelli, A., Sorrentino, N.: Energy management system for an energy district with demand response availability. IEEE Trans. Smart Grid **5** (5), 2385–2393 (2014)
5. Patti, E., Ronzino, A., Osello, A., Verda, V., Acquaviva, A., Macii, E.: District information modeling and energy management. IT Prof. **17**(6), 28–34 (2015)
6. Yuce, B., Rezgui, Y., Mourshed, M.: ANN-GA smart appliance scheduling for optimized energy management in the domestic sector. Energy Build. **111**(1), 311–325 (2016)
7. Yuce, B., Li, H., Rezgui, Y., Petri, I., Jayan, B., Yang, C.: Utilizing artificial neural network to predict energy consumption and thermal comfort level: an indoor swimming pool case study. Energy Build. **80**, 45–56 (2014)
8. Yang, C., Li, H., Rezgui, Y., Petri, I., Yuce, B., Chen, B., Jayan, B.: High throughput computing based distributed genetic algorithm for building energy consumption optimization. Energy Build. **76**, 92–101 (2014)
9. Yuce, B., Rezgui, Y.: An ANN-GA semantic rule-based system to reduce the gap between predicted and actual energy consumption in buildings. IEEE Trans. Autom. Sci. Eng. **PP** (99), 1–13 (2015)
10. Dibley, M.J., Li, H., Rezgui, Y., Miles, J.C.: An ontology framework for intelligent sensor-based building monitoring. Autom. Constr. **28**, 1–14 (2012)
11. Fanti, M.P., Mangini, A.M., Roccotelli, M., Ukovich, W., Pizzuti, S.: A control strategy for district energy managemet. In: 2015 IEEE International Conference on Automation Science and Engineering (CASE), Gothenburg, pp. 432–437 (2015)
12. Kalogirou, S.A.: Application of artificial neural network for energy systems. Appl. Energy **67**(1–2), 17–35 (2000)
13. Ferreira, P.M., Ruano, A.E., Silva, S., Conceicao, E.Z.E.: Neural networks based predictive control for thermal comfort and energy savings in public buildings. Energy Build. **55**, 238–251 (2012)

14. Gonzalez, P.A., Zamarreno, J.M.: Prediction of hourly energy consumption in buildings based on a feedback artificial neural network. Energy Build. **37**, 595–601 (2005)
15. Valerio, A., Giuseppe, M., Gianluca, G., Alessandro, Q., Borean, C.: Intelligent systems for energy prosumer buildings at district level. In: 23rd International Conference on Electricity Distribution (CIRED), pp. 1–5 (2015)
16. ISSDA, CER Smart Metering Project. http://www.ucd.ie/issda/data/commissionforen ergyregulationcer/. Accessed 15 Feb 2016
17. Gul, M.S., Patidar, S.: Understanding the energy consumption and occupancy of a multi-purpose academic building. Energy Build. **87**(1), 155–165 (2015)

Future Demand Response Services
for Blocks of Buildings

Tracey Crosbie[1(✉)], Vladimir Vukovic[1], Michael Short[1],
Nashwan Dawood[1], Richard Charlesworth[2], and Paul Brodrick[2]

[1] School of Science and Engineering, Teesside University, Middlesbrough, UK
{T.Crosbie,V.Vukovic,M.Short,N.Dawood}@tees.ac.uk
[2] Energy Management Division RC-GB EM DG PTI, Siemens plc,
Princess Road, Manchester, UK
{Richard.Charlesworth,Paul.Brodrick}@siemens.com

Abstract. Research surrounding demand response (DR) is beginning to consider how blocks of buildings can operate collectively within energy networks. DR at the level of a block of buildings involves near real-time optimisation of energy demand, storage and supply (including self-production) using intelligent energy management systems with the objective of reducing the difference between peak-power demand and minimum night-time demand, thus reducing costs and greenhouse gas emissions. To enable this it will be necessary to integrate and augment the telemetry and control technologies embedded in current building management systems and identify potential revenue sources: both of which vary according to local and national contexts. This paper discusses how DR in blocks of buildings might be achieved. The ideas proposed are based on a current EU funded collaborative research project called "Demand Response in Blocks of Buildings" (DR-BOB), and are envisaged to act as a starting-point for future research and innovation.

Keywords: Demand Response (DR) · Smart electricity networks · Micro-grids · Blocks of buildings

1 Introduction

Demand response (DR) provides an opportunity for consumers to play a significant role in the operation of the electric grid by reducing or shifting their electricity usage during peak periods in response to time-based tariffs or other forms of financial incentives. DR is widely recognised as being beneficial for customers, smart energy networks and the environment [1–3]. Specifically DR offers a number of benefits to energy systems including:

- Increased efficiency of asset utilisation;
- Supporting increased penetration of renewable energy on national energy grids;
- Easing capacity issues on distribution networks to facilitate further uptake of distributed generation on congested local networks;
- Reducing required generator margins and costs of calling on traditional spinning reserve;
- Bringing associated environmental benefits through reduced emissions.

© ICST Institute for Computer Sciences, Social Informatics and Telecommunications Engineering 2017
J. Hu et al. (Eds.): SmartGIFT 2016, LNICST 175, pp. 118–135, 2017.
DOI: 10.1007/978-3-319-47729-9_13

Research surrounding DR is beginning to consider how blocks of buildings can operate collectively within energy networks [4]. However, the majority of DR implementations aimed at small or medium scale customers have failed to meet their expected potential [5]. In this sense the value chain of DR service provision in blocks of buildings for the different actors involved has yet to be demonstrated [3]. These actors include but are not restricted to Distribution Network Operators (DNOs) Energy Retailers, Transmission Service Operators (TSOs), Energy Service Companies (ESCOs), IT providers, Aggregators[1] and facilities owners and managers.

The potential value of DR service provision in blocks of buildings depends on the telemetry and control technologies embedded in the building management systems currently deployed at any given site and the potential revenue sources: both of which vary according to specific local and national conditions [3]. In this context, to encourage the growth of DR services' and reap the potential benefits of DR, it is necessary for current research to demonstrate the economic and environmental benefits of DR for the different key actors required to bring DR services in blocks of buildings to market. This is the aim of a current EU funded project called "Demand Response in Blocks of Buildings" (DR-BOB) which is co-funded by the EU's Horizon 2020 framework programme for research and innovation.

To demonstrate the economic and environmental benefits of DR for the different key actors required to bring DR services in blocks of buildings to market the DR-BOB project has the following ambitious but achievable objectives:

- Integrating existing technologies to form the DR-BOB DR energy management solution for blocks of buildings with a potential Return on Investment (ROI) of 5 years or less.
- Piloting the DR-BOB energy management solution at 4 sites operating under different energy market and climatic conditions in the UK, France, Italy and Romania with blocks of buildings covering a total of 274,665 m^2, a total of 47,600 occupants over a period of at least 12 months.
- Realising up to 11% saving in energy demand, up to 35% saving in electricity demand and a 30% reduction in the difference between peak power demand and minimum night time demand for building owners and facilities managers at the demonstration sites.
- Providing and validating a method of assessing at least 3 levels of technology readiness (1-no capability, 2-some capability, 3-full capability) related to the technologies required for consumers' facilities managers, buildings and the local energy infrastructure to participate in the DR energy management solution at any given site[2].

---

[1] DR aggregation service providers are beginning to emerge in some EU energy markets. In Explicit Demand Response schemes (sometimes called "incentive-based") the aggregated demand side resources are traded in the wholesale, balancing, and capacity markets by energy aggregation service providers [6].

[2] The DR-BOB project's aims and objectives are detailed in the DR-BOB Grant Agreement (No 696114) and as such they are also listed on the European Commission's Community Research and Developments Information Service (CORDIS).

- Identifying revenue sources with at least a 5% profit margin to underpin business models for each of the different types of stakeholders required to bring DR in the blocks of buildings to market in different local and national contexts.
- Engaging with at least 2,000 companies[3] involved in the supply chain for DR in blocks of buildings across the EU to disseminate the project goals and findings.

2 Moving Beyond the State of the Art

There is a lack of integrated tools supporting optimisation, planning and control/management of supply side equipment [5]. As such, the majority of demand response implementations aimed at small or medium scale customers have failed to meet their expected potential [5] largely due to a lack of:

1. Relevant real-time information reaching customers from utilities due to outdated metering technologies and/or undue complexity in the presentation of information;
2. Means and abilities for customers to respond to real-time prices and demand signals, and few real incentives for them to do so;
3. Scalable integrated tools supporting optimisation, planning and control/management of supply side equipment which helps to perpetuate the energy industries general assumption of demand inelasticity [6–8].

The assembly of existing technologies, software components and concepts into a scalable, low cost and open optimisation platform for supply/demand optimisation and its evaluation in four representative demonstrations in EU member states are the key technical innovations in the DR-BOB project. Therefore the innovation in the project lies, not in the development of new technologies but rather in the integration of existing technologies and their application for DR at the level of blocks of buildings. The approach adopted will advance the state-of-the-art and address the barriers to DR for medium scale customers through:

1. The application of compact and efficient optimisation models to fully integrate supply and demand side optimisation for de-centralised neighbourhood scale power networks involving blocks of buildings and micro-grids;
2. The assembly and testing of a low-cost and mostly open source implementation platform for de-centralised DR in blocks of buildings, micro-grids and other neighbourhood scale power networks;
3. The configuration and augmentation of existing technologies to provide simple and effective user interfaces to encourage effective DR in decentralised neighbourhood scale power networks involving blocks of buildings and micro-grids.

---

[3] This figure while ambitious is based on the DR-BOB project partners experience of what is achievable during projects of this type.

3   Demand Response Energy Management Solution

The key functionality of the DR-BOB DR energy management solution (see Fig. 1) is based on the real-time optimisation of the local energy production, consumption and storage. It is envisaged that solution will be operated by an ESCO or energy management company/energy management department within an organisation. The criteria for the optimisation will be adjusted to either maximise economic profit or to minimise CO_2 emissions according to the requirements of the user. The energy management solution is intelligent in the sense that it is automated and can adapt to fluctuations in the energy demand or production, subject to dynamic price tariffs and changing weather conditions.

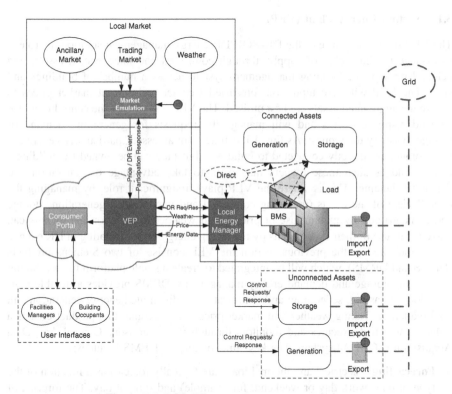

Fig. 1. DR-BOB DR energy management solution

The DR-BOB DR energy management solution will provide an innovative scalable cloud based central management system, supported by a local real-time energy management solution which communicates with individual building management systems and generation/storage solutions within a block of buildings. This will be achieved by integrating the following three tools and technologies:

- Virtual Energy Plant (VEP) – Siemens DEMS® [9] & Siemens DRMS [10]
- Local Energy Manager (LEM) –Teesside University IDEAS project Product [11]
- Consumer Portal – GridPocket EcoTroks™ [12]

The VEP will provide macro-level optimised energy management and the LEM will provide micro level optimised energy management, while the EcoTroks™ Customer Portal will provide user interfaces for energy management and community engagement. The solution will be applicable at all voltage levels, directly with low voltage (LV) and medium voltage (MV), depending on the sites at which it is deployed, in the case of high voltage (HV) the applicability will be indirect as many DR requests are sourced from the Transmission Network Operator (TSO).

3.1 Virtual Energy Plant (VEP)

The VEP sits at the centre of the DR-BOB DR energy management solution; its role is to manage the balancing of supply, through connected and unconnected assets, and demand through the building management system across a number of buildings and sites. The VEP will take inputs on forecasted weather and current market prices to ensure the aggregated assets are best utilised. The VEP will also be the central point for demand response events and will manage the requests for generation, demand or storage to satisfy the request. As Fig. 1 illustrates, not all assets, particularly generation and storage, are directly connected to buildings, but they can be owned by building's owners and as such these owners may want to take advantage of generation and demand balancing. This is where the VEP plays a significant role, by managing the overall view of the assets (including connected and unconnected generation, storage and demand) and by making use of pricing and weather information the owners can make better use of their investments by managing energy or participating in DR events.

As indicated in the previous section the VEP consists of two Siemens products (DEMS and DRMS), which will be integrated to create a single platform for combining generation, storage and demand energy management. DEMS provides a flexible platform for forecasting, scheduling and optimisation of distributed energy generation and load reduction. Taking weather and market prices into account, this product creates a schedule which it executes and monitors through its Supervisory Control And Data Acquisition (SCADA) system. The key functionality of DEMS includes:

- **Forecasting:** Electrical and thermal loads are typically forecast as a function of the type of day (work day or weekend, for example) and time of day. The forecast of renewable energy generation is also important, and is based on the weather forecast and the characteristics of the power plants. With parameterisable forecast bandwidth, it is possible to determine the reserve and risk strategies for plant operation in advance.
- **Planning:** Short-term scheduling for all the configured units is carried out in order to minimise the costs of power generation and plant operation in accordance with the general technical conditions and the terms of contracts. This is done in a 15 min time grid for up to a week in advance. The calculated dispatch plan minimises generation and operating costs. DEMS takes both economic and ecological factors

into consideration and can accommodate complex energy-supply/purchase contracts with power-zoned energy prices, time-dependent tariff structures, power bands, and energy limits.

- **Optimisation:** The optimised dispatch plan for thermal power plants takes into account power-up costs, maximum output ramps, minimum operating and shut-down times, fuel quantity limits, and energy limits, as well as time-dependent fuel prices. With regard to energy demands, equipment dispatch planning differentiates between three types of loads: independent loads, switchable loads, and controllable loads. Storage systems are managed according to specific user requirements. Real-time optimisation can be achieved based on the dispatch plan, any deviations are distributed cyclically at minimum cost among generators, storage systems, and loads, so that the planned value can be met. In this way, any external stipulations relating to import, supply, or corresponding contracts can be fulfilled.

DRMS (see Figs. 2 and 3) allows utilities and ESCOs to manage all aspects of their DR programs through a single, integrated system which provides an automated, integrated, and flexible DR solution. DRMS compliments DEMS by providing load reduction through surgical DR as it provides the ability to target, a single asset or all assets in an area, or all assets attached to a given network device (such as a feeder or transformer).

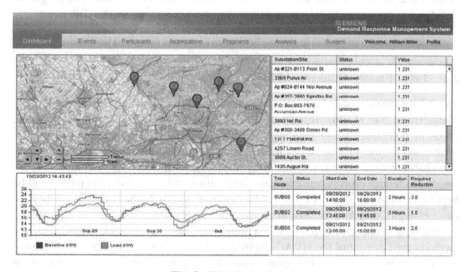

Fig. 2. DRMS dashboard

- DR capacity can be cost effectively scaled by automating the manual processes that are typically used to execute DR events and settlement.
- DR resources can be used in a more intelligent and efficient way by planning and executing load shed at grid locations where the utility has more benefit.

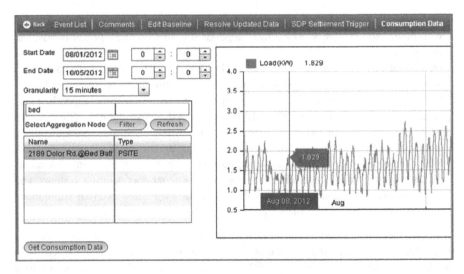

Fig. 3. DRMS consumption data screen

DRMS can be configured to support multiple types of DR programs providing both emergency and economic dispatch. DRMS can interface with residential, commercial, or industrial sites to provide more flexibility in how ESCOs create their DR programs. DRMS also provides the ability to define workflow processes so that DR events are managed according to utility business processes.

3.2   Local Energy Manager (LEM)

The VEP capability will be further enhanced by the introduction of a Local Energy Management tool (LEM). Instead of the VEP having to scale to meet the real-time needs of potentially many buildings and their specific energy management requirements; and for the VEP to be continually updated in real-time with status information; the VEP will hand off this responsibility to individual LEMs. This will improve reliability and reduce costs because the communications and centralised computing will be kept to a minimum. As the LEMs will be deployed locally to the buildings (if they are close enough), they will be able to manage the energy usage much more effectively.

Building management has come a long way since it was first introduced, the methods used to control the HVAC, lighting and other energy consumers is becoming much more sophisticated. However, when looking at a DR Energy Management Solution such as that discussed here, where some of the buildings, and the Building Management System (BMS) solutions they have deployed, are becoming antiquated in their capabilities and unable to manage the addition of generation and storage assets, the LEM plays an important role in the energy management of the building by providing the additional capabilities needed to optimise the operation. The LEM will be based upon a rack-mounted industrial server supplied by Siemens and will be equipped with heat and electrical load prediction and compact commitment/dispatch optimisation

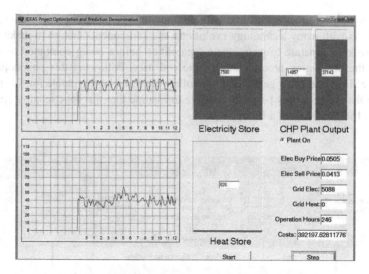

Fig. 4. IDEAS neighbourhood energy optimisation and prediction tool

software developed previously in the IDEAS project. Figure 4 presents a screen shot of the IDEAS tool demo version.

The LEM will integrate with the local generation and storage assets and BMSs to manage demand. This integration will enable the LEM to manage the assets locally by taking a local view of the buildings environment. Also, where assets aren't directly connected to the BMS, which is often the case with generation, the LEM will be able to integrate directly. By taking this local view the LEM will be able to take the responsibility for the decisions on what load, storage or generation it has to spare, informing proactively, or on-demand, the VEP of its status so that DR events can be dispatched appropriately. Technically, the marginal prices for shifting and shedding load will be regularly communicated in the form of contract options to the VEP which acts to reduce peak loads. Following the occurrence of a specific DR event in the local market, the LEMs may also resolve the local commitment/dispatch optimisation problem subject to additional inequality/equality constraints to recover a specific price for implementing the DR request, which is communicated back to the VEP. This is a form of two tiered decentralised optimisation which is known to be effective in smart-grid applications [13]. In addition to reducing computational demands upon the VEP, the proposed integration with the LEM will increase security, as detailed internal operational information related to a building does not need to leave the domain managed by LEM.

The optimisation embedded in the LEM will build on work conducted as part of the IDEAS project.[4] The generic approach shown in the Fig. 5 illustrates how the real-world data and predictions will be used for optimisation and decision support. It is

[4] IDEAS Collaborative Project (Grant Agreement No. 600071) which was co-funded by the European Commission, Information Society and Media Directorate-General, under the Seventh Framework Programme (FP7), Cooperation theme three, "Information and Communication Technologies", http://www.ideasproject.eu.

planned to handle direct automation with building energy infrastructures via interfaces to BMS. As with many technology evolutions, the components that go to make up the whole are kept separate to prove their capabilities before being combined to reduce costs in the production process and for the end user. Following this philosophy the LEM, the BMS and other generation controllers are to be kept separate for this project to prove the concept; it is perfectly feasible to assume that this functionality could be embedded in the BMS at a future stage.

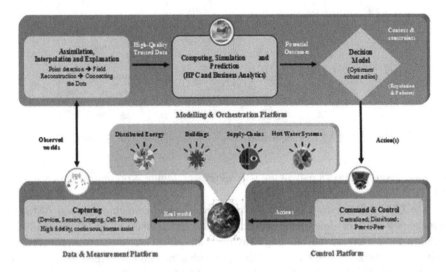

Fig. 5. Optimisation and decision support architecture

3.3   EcoTroks™ Customer Portal

The EcoTroks™ application is multi-device friendly and based on unified platform for all web browsers, including tactile web terminals [12]. It has several customisable features and extendable widgets and supports multiple languages. Essentially it will provide a customer portal (see Fig. 6) with an online guide and stimulation tool for energy users. EcoTroks™ can be used to engage users in DR using functionalities based on behavioural theories designed to improve smart grid stability and performance. EcoTroks™ motivates people to reduce energy consumption, use DR tools and shift consumption to off-peak hours by proposing gamification scenarios, personal and group challenges, auto-reflexive consumption visualisation and contact with relevant communities. It helps users to understand the impact of actions on energy consumption via personalised interfaces, accessibility of EcoTips and achieves sustainable change by motivating users via virtual currency.

Consumption analysis features include:

- Analysis of daily consumption graphics;
- Historical daily, monthly, weekly, yearly consumption insights;
- Real-time consumption analysis;

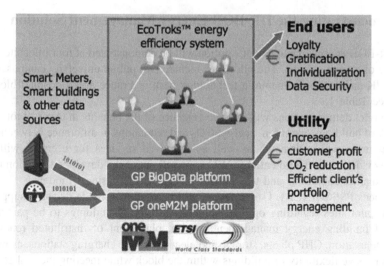

Fig. 6. GridPocket EcoTroks™ platform overview

- High level data disaggregation (air-conditioning, heating system, hot water);
- Peak/Off peak hours consumption control;
- Actual energy costs and spending (support multiple tariffs);
- Consumption prediction for performance analysis.

3.4 Systems Configuration

The configuration of the DR-BOB energy management solution will allow energy management companies to provide varying levels of control from the centralised macro-view, through to localised complete control of the energy systems at the building level, the micro-view. The solution will utilise existing standards such as IEC60870-5-104 and OpenADR, and an architecture that will enable new adaptors to be added to support new standards in the future. These standards allow access to most generation, storage and load assets. It is expected that any new interfaces between the platform and the ESCO could form the basis for new standards.

In combination, DR-BOB DR solution will provide open connectivity to both SCADA/utility communications and customer side advanced metering infrastructures. The decentralised approach – allowing both supply side and DR to be hierarchically optimised between blocks of buildings and other infrastructures, with automatic distribution of results via building management systems – removing some of the burden and alleviating the complexities involved in individual customer or resident participation.

The advantage of Siemens products in this solution is their support for standards such as IEC61879-5-104 and OpenADR, and flexible architecture that will enable new adaptors to be added to support new standards in the future. These standards allow access to most generation, storage and load assets. It is expected that any new interfaces between the platform and the ESCO could form the basis for new standards.

4   Demonstrating the DR-BOB Energy Management Solution

The DR-BOB energy management solution will be demonstrated at four pilot sites over a period of twelve months. The pilot sites include two public university campuses one in the UK and one in Romania, a technology park in France and a hospital block in Italy (see Table 1).

The pilot demonstrations will provide evidence of the benefits in terms of total cost of block of buildings operation, energy, CO_2 and reductions in difference between peak and minimum power demand, as well as improved services for users; it will also provide evidence for the potential for EU wide roll out of the developed solution under diverse market conditions and test various aspects of the business models.

Siemens PSS®SINCAL [16] software will be used to provide decision support at the four pilot sites regarding optimal number and type of buildings to be part of the block of building energy management system, placement of distributed renewable energy generation, CHP plants, storage and electric vehicle charging stations, as well as built-in passive flexibility of buildings within the block while meeting thermal comfort requirements. PSS®SINCAL is a mature technology that is used by over 500 organisations worldwide, including some 300 in Europe. PSS®SINCAL provides a full unbalanced power system model for high, medium, and low-voltage grids and supports the design, modelling and analysis of electrical power systems, as well as pipe networks, such as water, gas, and district heating/cooling systems. Through its modular and fully integrated design PSS®SINCAL enables a high level of customisation according to individual needs, making it the optimal solution for all planning tasks in the areas of generation, transmission, distribution and industrial grids.

The demonstrations will be conducted in three main phases:

- The first stage will investigate application and acceptance of the DR measures at the demonstration site in the UK, including market analysis and financial implications.
- The second stage will involve the implementation and monitoring of the cloud based infrastructure performance at the other three demonstration sites in France, Italy and Romania, including the decision support tool for control management, and user interfaces at all four demonstration sites over a period of one year.
- The third stage will develop and evaluate EU wide deployment strategies and business models to fit different energy markets operating across Europe. It will illustrate the value proposition underpinning the business cases for all stakeholders and discuss specific investment strategies.

5   Demand Response Markets

The ability to realise the benefits of DR is dependent on the market structure and the way in which varies across the different countries in the EU. In the case of the utilities the business value to be gained from investment in DR in part, is dependent upon how the supply, distribution, transmission and generation functions of the utilities industry are distributed between the different actors in the energy supply chain. This varies across the EU depending upon the degree to which, and way in which, different EU countries have unbundled these traditionally vertically integrated markets. Essentially

Table 1. DR-BOB pilot site characteristics

Site	Buildings	Technologies	Climate[a]	Market[b]	Targets
UK	Educational, office, catering + low rise residential	CHP, EV charging stations, RES (PV)	Temperate oceanic (Cfb)	British Isles	−17% el. demand
					−7% en. demand
					−30% peak-min d.
					ROI 5 yrs.
FR	Workshop, training centre, office	Microgrid, EV charging stations, RES (PV), electric storage	Temperate oceanic (Cfb)	Central Western Europe	−11% el. demand
					−11% en. demand
					−30% peak-min d.
					ROI 5 yrs.
IT	Healthcare + office	RES (PV), thermal storage, CCHP (trigeneration), DH	Humid subtropical (Cfa)	Apennine Peninsula	−9% el. demand
					−6% en. demand
					−30% peak-min d.
					ROI 1 yr.
RO	Educational, leisure, office + high rise residential	RES (PV, wind), thermal storage, CHP	Temperate/humid continental (Dfb)	Central Eastern Europe	−35% el. demand
					−10% en. demand
					−30% peak-min d.
					ROI 3.9 yrs.
Total	**274,665 m²**	**8 different technologies**	**Representative of 61% EU**	**4 out of 7 EU markets**	**−8% en. demand**
	47,600 occupants				**−21% el. demand**
					−30% peak-min d.
					ROI < 5 yrs.

[a]According to Koppen climate classification [13]
[b]Regional wholesale electricity markets [14]

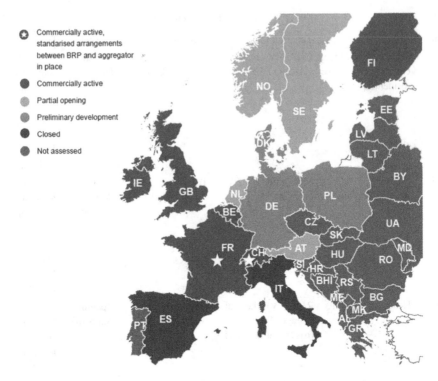

Fig. 7. Demand response map of Europe 2014 (Source [6])

the way in which the functions of the utilities are separated affects the impetus to promote integrated resource planning and therefore decreases the value proposition underpinning load management initiatives [17]. This is illustrated by research, which indicates that the way in which the utilities industries have adapted their corporate structures to adhere to the EU directives for the unbundling of the utilities industry affects the value creation available from renewable energy generation and investment in smart metering and smart networks [18]. These findings are further emphasised by research which highlights that the liberalisation of the utilities industries can have conflicting consequences for the implementation of demand side management within energy supply and sustainable business practices in general [19].

Some European energy markets are more mature in terms of their support for demand side management than others: particularly in relation to the introduction of variable and time of use tariffs [4]. Nevertheless, a market for DR is emerging in some EU countries [6]. Six European countries already provide a regulatory framework allowing for the development of DR services: Ireland, Great Britain, Belgium, France, Switzerland and Finland (see Fig. 7). Although there are remaining regulatory issues. Explicit DR is a commercially viable product offering [6]. In the UK for example the potential revenue sources for Energy Retailers, Distribution Network Operators (DNO) and Transmission Systems Operators (TSO) are growing (see Fig. 8 and Table 2).

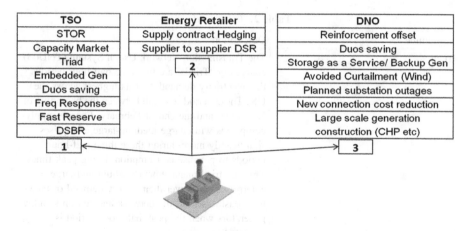

TSO	Energy Retailer	DNO
STOR	Supply contract Hedging	Reinforcement offset
Capacity Market	Supplier to supplier DSR	Duos saving
Triad	2	Storage as a Service/ Backup Gen
Embedded Gen		Avoided Curtailment (Wind)
Duos saving		Planned substation outages
Freq Response		New connection cost reduction
Fast Reserve		Large scale generation
DSBR		construction (CHP etc)
1		3

Fig. 8. Contract models – new and pending – where an asset could participate (Model developed by Siemens from literature and conversations with regulator and market participants.)

To address the uneven development of markets for DR in blocks of buildings the DR-BOB project will seek to identify how mechanisms for DR from more mature markets could be implemented in EU countries with less mature markets for DR. The results will provide feedback to the market participants with recommendations on how that country could adopt a new mechanism and what value there is in doing so for the different actors involved in the value chain required to bring DR in blocks of buildings to market.

Table 2. Available UK contract model descriptions

Programme	Description
UK National Grid Short Term Operating Reserves Programme (STOR)	This programme is part of National Grid's Balancing Mechanism. The programme is open to both supply side and demand side generators. The primary purpose of the programme is to provide frequency regulation in response to short term interruptions to supply.
Capacity market	The UK government has introduced a capacity market to ensure the electricity system has adequate supplies available in the future. The available capacity payment may make it economical to build small, quickly despatchable distributed generators without needing to rely on energy market revenues.

(continued)

Table 2. (*continued*)

Programme	Description
Triads	The Transmission Network Use of System (TNUoS) charges or 'Triads' are the three half-hour periods that electricity demand is at its highest across the UK. These periods can fall between the beginning of November and the end of February. Energy supply companies will charge medium-large enterprises significantly more during these three half-hour periods to penalise consumption during peak times. Through triad management medium and large enterprises can lower their energy demand or take their businesses off the network and run on standby generators when the peak half-hour period is expected to occur.
Distribution Use of Systems (DUoS)	DUoS is a time of use tariff similar to triads. However, DUoS happens on a daily basis and is added onto the monthly electricity bill. It is the charge for receiving electricity from the national transmission system and transferring it to the distribution level to be used in homes and businesses. DNOs charge energy supply companies, such as British Gas, E.ON, and others, who then pass these charges onto electricity bills to cover the costs of installing, operating and maintaining the network. DUoS can account for up to 12% of the charges on an electricity bill for half-hour meter industrial and commercial customers.
Frequency response	System frequency is a continuously changing variable that is determined and controlled by the second-by-second (real-time) balance between system demand and total generation. If demand is greater than generation, the frequency falls while if generation is greater than demand, the frequency rises. The UK TSO, National Grid, will pay a premium for participation in such events.
Fast reserve	Active power delivery must start within 2 min of the despatch instruction at a delivery rate in excess of 25 MW/minute, and the reserve energy should be sustainable for a minimum of 15 min. Must be able to deliver minimum of 50 MW. Fast Reserve provides the rapid and reliable delivery of active power through an increased output from generation or a reduction in consumption from demand sources, following receipt of an electronic despatch instruction from National Grid.

(*continued*)

Table 2. (*continued*)

Programme	Description
Retail supply side contract hedging	Retailer specific demand response includes: being able to optimise the market based transactions that can have the ability to help realise significant savings being utilised as a hedge against their supply-side contracts, or generate maximum revenues at times of high demand and increased supply prices. Essentially suppliers will use demand side response (DSR) within-day (period between day-ahead and gate closure) to re-align their positions.
DNO Traditional Network Reinforcement Offset	Used by local network to avoid or defer network reinforcement. The DNO will use DSR to tackle planned outages and unplanned outages as well as critical peak scenarios. Requirements for planned outages are generally known at least one day in advance. For unplanned outages, DSR will need to be called sufficiently quickly to prevent a circuit trip or risk of unacceptable loss of asset life due to thermal stress on network components. For subsequent outage days, DSR units may have 24 h of notice.
Avoided curtailment	Increase demand to avoid wind curtailment or to soak up solar output. These issues are intrinsically related to the management of intermittent generation (intermittency Management). Suppliers (vertically integrated entities), the Service Operator (SO) or even wind portfolio players may wish to increase demand to avoid wind/solar curtailment or reduce demand to mitigate the effects of low wind periods (low wind periods typically coincide with peak price periods especially in the winter). In addition, the SO may wish to use DSR to reduce the level of peak generation capacity needed on the system. This will be incentivised through the capacity payment.

6 Conclusion

The assembly of existing technologies, software components and concepts into a scalable, low cost and open optimisation platform for supply/demand optimisation and its evaluation in four representative demonstrations in EU member states are the key technical innovations in the DR-BOB project. The proposed work is ambitious in that specific solutions to identified problems and barriers to effective DR are intended to be developed, but also realistic in its stated aims.

The individual components required for the DR-BOB DR energy management solution are already in existence and in some cases mature. For example, open communication architectures, protocols and standards definitions (e.g. IEC 60870-5/6, OpenADR) and efficient open-source Mixed Integer Linear Programming (MILP) tools such as LPsolve are in widespread use. In addition, several proposals have been made for decentralised supply/demand optimisation in micro-grid environments using agent-based techniques. However the effectiveness and reliability of these latter techniques has not been extensively demonstrated in real-test systems', and this is seen as a key impediment to their widespread use. The approach presented in this paper seeks to remove that impediment. The DR-BOB project will run from March 2016 until February 2019 and the first results of the project are due in 2017.

Acknowledgement. The work presented was carried out as part of the DR-BOB project (01/03/16 - 28/02/19) which is co-funded by the EU's Horizon 2020 framework programme for research and innovation under grant agreement No 696114. The authors wish to acknowledge the European Commission for their support. The authors also gratefully acknowledge the contribution of the DR-BOB project partners to the work presented from: Siemens, Teesside University, Centre Scientifique et Technique du Bâtiment; R2M Solutions; Nobatek, Gridpocket; Duneworks; Fondazione Poliambulanza; Servelect and the Universitatea Tehnica Cluj-Napoca.

References

1. Smart Energy Demand Coalition: A Demand Response Action Plan for Europe (2013). http://smartenergydemand.eu/wp-content/uploads/2013/06/SEDC-DR-FINAL-.pdf
2. Bradley, P., Leach, M., Torriti, J.: A review of current and future costs and benefits of demand response for electricity. Centre for Environmental Strategy Working Paper, vol. 10, no. 11 (2011)
3. Thomas, C., Star, A., Kim, J.: An assessment of business models for demand response. In: Proceedings of Grid-Interop Forum, Denver (2009)
4. Crosbie, T., Dawood, M., Short, M., Brassier, P., Dorcome R., Huovila, A., Ala-Juusela, M.: IDEAS Project Deliverable D2.3, Generalised business models (2015). https://www.researchgate.net/publication/287948614_Generalised_business_models
5. Olivares, O., Mehrizi-Sani, A., Etemadi, A.H., Cañizares, A.H., Iravani, R., Kazerani, M., Hajimiragha, A.H., Gomis-Bellmunt, O., Saeedifard, M., Palma-Behnke, R., Jiménez-Estévez, G.A., Hatziargyriou, N.D.: Trends in microgrid control. IEEE Trans. Smart Grid **5** (4), 1905–1919 (2014)
6. Smart Energy Demand Coalition: Mapping demand response in Europe today (2015). http://www.smartenergydemand.eu/wp-content/uploads/2015/09/Mapping-Demand-Response-in-Europe-Today-2015.pdf
7. Torriti, J., Hassan, M.G., Leach, M.: Demand response experience in Europe: policies, programmes and implementation (2010)
8. Kim, J.H., Shcherbakova, A.: Common failures of demand response. Energy **36**, 873–880 (2011)
9. Siemens: Virtual Power Plants by Siemens, DEMs-Decentralizied Energy Management System (2013). http://w3.usa.siemens.com/smartgrid/us/en/distributech/Documents/DEMS_VPPs.pdf

10. Siemens: Siemens DRMS Demand Response Management System – Version 2.0 (2013). http://w3.usa.siemens.com/smartgrid/us/en/demand-response/demand-response-managem ent-system/Documents/DRMS_SellSheet_V2.pdf

11. Short, M., Dawood, M., Gras, D., et. al.: IDEAS Project Deliverable D4.1: A prototype neighbourhood energy management tool (2015)

12. GRIDPOCKET: EcoTroks™ energy efficiency. http://www.gridpocket.com/products/beha vioral-energy-efficiency-program/

13. Li, M., Luh, P.B.: A decentralized framework of unit commitment for future power markets. In: IEEE Power & Energy Society General Meeting (2013)

14. Peel, M.C., Finlayson, B.L., McMahon, T.A.: Updated world map of the Köppen-Geiger climate classification. Hydrol. Earth Syst. Sci. **11**, 1633–1644 (2007). doi:10.5194/hess-11-1633-2007. ISSN 1027-5606

15. DG Energy: Quarterly Report on European Electricity Markets, vol. 7, no. 3 (2014). http://ec.europa.eu/energy/sites/ener/files/documents/quarterlyelectricity_q3_2014_final_0.pdf

16. Siemens: PSS®SINCAL – efficient planning software for electricity and pipe networks, Siemens PTI – Software Solutions (2010). http://w3.siemens.co.uk/smartgrid/uk/en/Servi ces/as/brochures/Documents/PSS%C2%AESINCAL_efficient%20planning%20software.pdf

17. Kelly, A., Marvin, S.J.: Demand side management: the electricity sector in town planning in the UK. Land Use Policy **12**(3), 205–222 (1995)

18. Tselentis, C.: Innovation in European electricity utilities: an interdisciplinary approach. In: 3rd Annual Conference Competition and Regulation in Network industries 19th November, Residence Palace, Rue de la Loi 155, Brussels, Belgium (2010)

19. Crosbie, T.: The utilities in transition: gazing through the IT window. Flux **75**(1), 16–26 (2009). International Scientific Quarterly on Networks and Territories

Use Cases and Business Models of Multi-Agent System (MAS) ICT Solutions for LV Flexibility Management

Juan Manuel Espeche[1(✉)], Thomas Messervey[1], Zia Lennard[1], Riccardo Puglisi[1], Mario Sissini[2], and Meritxell Vinyals[3]

[1] R2M Solution, Pavia, Italy
{juan.espeche,thomas.messervey,zia.lennard,
riccardo.puglisi}@r2msolution.com
[2] Smart Metering Systems plc, Cardiff, UK
[3] CEA-LIST, Gif-sur-Yvette, France

Abstract. This paper describes the use cases and business models opportunities of a Multi-Agent System (MAS) ICT solution for LV Flexibility Management. The MAS platform provides a technological solution that enables new collaboration opportunities between actors in the LV portion of the grid, namely, distribution system operators, ESCOs (in particular Telecoms) and consumers/prosumers. MAS have potential for efficient decision-making in the LV part of the grid due to the large number devices, users and variables and which makes more efficient a decentralized decision making approach. To support the new collaborations and business strategies amongst these actors, new business models are required and the ecosystem forms series of multi-sided platform business models. In this paper, the approach to business model development is detailed and 17 resultant business model opportunities are identified. These business models are then mapped to the use cases for future analysis.

Keywords: Smart grid · Multi-sided platform business models · Flexibility management · Capacity management · Multi-agent system · Optimization · Aggregators

1 Introduction

The Mas[2]tering project is centered upon the development of a multi-agent system (MAS) ICT platform [1] for use by local flexibility aggregators to conduct the flexibility management of local energy communities in the low voltage (LV) portion of the grid. It is this part of the grid that is most changed by the transition to the smart grid and where decentralized decision making has the most potential to bring value and competitiveness. In addition, this ICT platform provides a technological solution (smart grid access point) to connect to the smart grid.

The research leading to these results has received funding from the EU Seventh Framework Programme under grant 619682 (Mas[2]tering).

© ICST Institute for Computer Sciences, Social Informatics and Telecommunications Engineering 2017
J. Hu et al. (Eds.): SmartGIFT 2016, LNICST 175, pp. 136–142, 2017.
DOI: 10.1007/978-3-319-47729-9_14

In the smart grid and in simple terms, flexibility can be defined as changes to the consumer or prosumer planned consumption, production or storage profiles. If aggregated and managed properly, flexibility can be a tool to increase grid efficiency, reduce peak demand and defer or avoid expensive grid reinforcements. In the USA, one study estimates that residential flexibility can avoid $9 billion in planned grid investments, avoid $4 billion in energy production and ancillary service costs and reduce consumer energy bills by between 10–40% [3]. In Europe, 2015 communications from the European Commission call for a consultation for energy market redesign [4] and new deal for Europe's energy consumers [5]. In both communications, flexibility concepts and access to participate in the smart grid are central themes and a clear need to address capacity management is cited as the grid's most pressing challenge (e.g. the need for solutions to answer increased electrification, increased urbanization, and the increased connection of intermittent renewable energy technologies to the low and medium voltage parts of the distribution grid infrastructure). Mas^{2}tering provides solutions to these challenges using a multi-agent system approach to optimize the shift of controllable loads from prosumers in the LV grid in order to achieve a set of goals (reduce energy bill, maximize self-consumption, optimize efficiency of a local energy community, optimize efficiency of a local area of the grid, reduce LV congestion, increase grid reliability, and others). The solution connects consumers to the grid in a new way, facilitates the aggregator role, and provides a service to DSOs for capacity and congestion management.

Equally important as the technological solutions are the market model framework, interactions between market participants and business models to make the system viable. In this paper, the project use cases and initial business model opportunities are detailed. The use cases expand upon a progressive project vision and storyline by which:

- Consumers/prosumers lower their individual energy bills by using flexibility to maximize self-consumption and to adapt consumption to variable tariff schemes
- Consumers/prosumers forming local energy communities (as islands) where flexibility can be managed to lower the energy bills of the community and also to create benefits to the grid
- Local energy communities considered with grid constraints where Distribution System Operator (DSO) capacity management, other local energy communities or the flexibility market at large may create flexibility requests.

The primary actors involved in enabling residential flexibility are utilities, DSOs, telecoms and local flexibility aggregators. Together with consumers and prosumers they form a multi-sided platform with the value flows of electricity, data and revenue. This ecosystem and the resultant multi-sided business models are consistent with recent advances in smart grid business model development work [6].

Through literature review, stakeholder consultation and following the approach detailed in Sect. 3, 17 individual primary and secondary business model opportunities are identified and mapped to the project uses cases. They are clustered by (*Primary*): flexibility as a product, in-home optimization services, flexibility services to the DSO, joint services business models, and (*Supporting*): knowledge and data services, telecom services, security services and referral services. These business model opportunities will later be constructed into actor specific multi-sided platform business models.

2 Use Cases of Multi-agent System ICT Solutions for LV Flexibility Management

Three use cases are used to develop and validate project results. Each use case deals with a portion of the LV grid and has specific general and quantified objectives. Each use case is an enabler to the subsequent one as each analysis deals with a portion of the LV distribution network that is gradually wider than the previous one. The project use cases are:

Use Case 1: This UC focuses on the Home Area Network and the services that involve the LV end user (domestic and small commercial); the scope is the interoperability between the HAN management system, the smart meter and a technical interface (gateway) which allows the bi-directional communication between the end user and the rest of the LV grid. The enabled communication is a prerequisite to the local optimization proposed in the other UCs and, for the prosumer, to enter the market of flexibility products.

Use Case 2: This UC focuses on the district, intended as a community of prosumers represented by a local aggregator in a local area of the LV grid; the scope is to demonstrate that MAS optimization performed at this local level is effective for energy management and local balancing, as an alternative to traditional centralised optimization. The objective is to maximise revenue for prosumers belonging to the local community when coping with variable external conditions and conflicting requests from the DSO and the other grid actors.

Use Case 3: This UC is an extension of UC2 and is the union of multiple districts in a given area (represented by a MV/LV substation). The UC targets in particular DSOs and aims at demonstrating that the local optimization enabled in UC2, coupled with proper grid monitoring can be a cost-effective way to deal with local congestions and globally increase grid performances, reliability and resilience.

3 Business Models of Multi-agent System ICT Solutions for LV Flexibility Management

The approach to construct business model opportunities was:

- Value Flow Analysis: In multi-sided business models an actor may both a buyer and a seller. Mapping the flows (data, energy, revenue) in the multi-sided business model and identifying what each actor's role might entail within each flow.
- Value Proposition Generation: The overall value proposition towards the buyer needs to be able to answer: What customer problem is being solved; what customer needs

are being satisfied; what segment-specific products and services can be offered to customers; and what value is generated for customers?

- Value Chain Delivery Analysis: The value chain behind the creation of the value flow needs to account for internal resources; activities and competencies; partners; and distribution channels.
- Value Chain Capture: The revenue model needs to answer: Principle costs are in the proposed business model; what are the proposed sources of revenue; what is the customer willing to pay for; how do customers pay at present; and how should they pay in the future?
- Constraint Analysis: The constraints in the delivery of each model will vary based on several factors from individual national regulations, market conditions and the stakeholders' willingness to collaborate and engage with the opportunity.

3.1 Business Model Opportunities

Following the methodology explained above and consulting the multi-disciplinary consortium of EU Mas²tering project (Telco, DSO, retailer, research centers, small and medium enterprises), this research has identified 17 business model opportunities to investigate surrounding district flexibility management services. These opportunities are separated into primary and supporting business models. Primary business models are those that directly relate to the grid efficiencies possible by unlocking consumer/prosumer flexibility. Supporting business models are those associated with entering or facilitating the ecosystem. The business model opportunities are intentionally disaggregated to facilitate the identification and consideration of market analysis aspects, strategy options and collaboration opportunities.

- *Primary Business Models*

Flexibility as a Product

- **B1. Sale of Flexibility by a consumer/prosumer to a Local Aggregator** is appropriate for consumers/prosumers that do not require services related to in-home optimization and desire to gain value from offering flexibility to the market through an aggregator.
- **B2. Sale of Flexibility by a Local Aggregator to the Flexibility Market** deals with flexibility that a local aggregator may sell for purposes other than DSO Services. This flexibility is made available to the flexibility market where the buyer may be a DSO who does not require services, an aggregator of aggregators, a BRP or other market participant.
- **B3. Sale of Flexibility by a Local Aggregator Service Contract to a predetermined buyer (DSO, Aggregator of Aggregators or BRP)** deals with the predetermined sale of flexibility to a contracted buyer.

Consumer: In home optimization services

- **B4. Time-of-use (ToU) optimization** is based on load shifting from high-price intervals to low-price intervals or even complete load shedding during periods of high prices.

- **B5. Self-balancing** is typical for Prosumers who also generate electricity (for example, through solar PV or CHP systems). Value is created through the difference in the prices of buying, generating, and selling electricity.
- **B6. Control of the maximum load** is based on reducing the maximum load (peak shaving) that the Prosumer consumes within a predefined duration (e.g., month, year), either through load shifting or shedding. By lowering maximum load the consumer benefits from lower tariffs.
- **B7. Bundled Flexibility Management Service** the combination of optimization services coupled with flexibility sales to a LFA.

DSO: Flexibility services for DSO

- **B8. Congestion management** deals with the use of flexibility to attain the benefits of peak reduction, local balancing, the reduction of losses and voltage management in a discrete timeframe of high demand to avoid the thermal overload of system components.
- **B9. Grid capacity management** deals with the use of flexibility to conduct congestion management but also in a longer-term horizon to defer grid investments to ensure future capacity needs and to extend the operational lifetime of system components.

Joint Services Business Models

- **B10. Bundled Contracts (Phone-Internet-Energy) for the providing of In-Home Optimization and Flexibility Management Services** deals with strategic alliances between utilities and telecoms to offer bundled services or with 3rd party organizations that self-organize to offer holistic bundled solutions.

- *Supporting Business Models*

Knowledge & Data Services

- **B11. The sale of congestion point forecasting to local aggregators as a service** deals with the ability to create and deliver a competitive advantage and work avoidance via forecasting services.
- **B12. The sale of consumer/prosumer consumption data to Local Aggregators or Common Reference Point Operators as a service** deals with the data flow concerning consumer/prosumer load profiles and/or flexibility potential.
- **B13. The sale of MAS IP to Local Aggregators to maximize price differentials between flexibility purchases and flexibility sales** deals with the business model for the exploitation of the MAS foreground as it relates to the ICT platform/MAS IP.
- **B14. The sale of MAS IP to In-Home Agent Manufacturers (white goods and renewable energy technologies) to increase product competitiveness and differentiation** deals with the exploitation of the foreground MAS IP.

Telecom Services

- **B15. Broadband Content, VAS and OTT sales** deals with the sale of content licensing, Value Added Services(VAS) or Over the Top Content (OTT) subscription-based services by Telecom Operators to combined energy suppliers/LFA for enhanced device abstraction interoperability within major smart appliance categories connected to ZigBee, Energy@Home, 5G, cloud access channels, etc.

- **B16. HAN Sales** deals with the provision of Smart Gateway and related products/services in the HAN. According to the specific country and type of market the Telco company may also be the owner of the device and ask the final user to pay a fixed rate.

Security Services

- **B17. The sale of security software to ensure the secure transport of consumer/ prosumer data** deals with how software and data security providers add and take value from the system.

3.2 Use Case/Business Model Opportunities Mapping

See Table 1.

Table 1. Business models mapping into use cases

Business model opportunities	UC1	UC2	UC3
B1. Sale of flexibility by a consumer/prosumer to a local aggregator	x	x	
B2. Sale of flexibility by a local aggregator to the flexibility market		x	
B3. Sale of flexibility by a local aggregator service contract to a predetermined buyer		x	x
B4. Time-of-use (ToU) optimization	x		
B5. Self-balancing	x		
B6. Control of the maximum load	x		
B7. Bundled flexibility management service	x		
B8. Congestion management			x
B9. Grid capacity management			x
B10.Bundled contracts (Phone-Internet-Energy) for the providing of in-home optimization and flexibility management services	x		
B11. The sale of congestion point load profile forecasting to local aggregators as a service			
B12. The sale of consumer/prosumer consumption data to local aggregators or common reference point operators as a service	x		
B13. The sale of MAS IP to local aggregators to maximize price differentials between flexibility purchases and flexibility sales		x	
B14. The sale of MAS IP to in-home agent manufacturers to increase product competitiveness and differentiation	x		
B15. Broadband Content, VAS and OTT sales	x		
B16. HAN Sales	x		
B17.The sale of security software to ensure the secure transport of consumer/prosumer and aggregate data	x	x	x

4   Conclusions

This paper presented the business model scope of a multi-agent system ICT platform for LV flexibility management in the form of 17 collaboration opportunities. There are 10 primary business models (including "Flexibility as a Product", "Consumer: In home optimization services", "DSO: Flexibility services", and "Joint Services Business Models"), and 9 supporting business models (Including "Knowledge & Data Services", "Telecom Services", Security Services", and "Referral Services). This extensive approach of business models is the first step to consolidate final business models through further studies. The main benefit of identifying the business models is to provide a competitive edge. Implementing unique business model can give companies a unique reputation in the marketplace, creating buzz among consumers and encouraging first-time purchases.

A mapping between the business model opportunities and the use cases is presented, which facilitates the alignment of business models and technical architectures and therefore the gap between the research and the market is reduced.

References

1. Mourshed, M., Robert, S., Ranalli, A., Messervey, T., Reforgiato, D., Contreau, R., Becue, A., Quinn, K., Rezgui, Y., Lennard, Z.: Smart grid futures: perspectives on the integration of energy and ICT services. Energy Procedia **75**, 1132–1137 (2015)
2. Juziuk, J.: Design patterns for multi-agent systems. Master's dissertation, Linnæus University, Småland, Sweden (2012)
3. Bronski, P., Dyson, M., Lehrman, M., Mandel, J., Morris, J., Palazzi, T., Ramirez, S., Touati, H.: The Economics of Demand Flexibility: How "flexiwatts" Create Quantifiable Value for Customers and the Grid. Rocky Mountain Institute, Boulder (2015)
4. EC: Launching the public consultation process on a new energy market design, COM (2015) 340. European Commission, Brussels, Belgium (2015)
5. EC: Communication from the commission to the European parliament, the council, the European economic and social committee and the committee of the regions Delivering a New Deal for Energy Consumers. EU COM(2015) 339. European Commission, Brussels, Belgium (2015)
6. Valocchi, M., Juliano, J., Schurr, A.: Evolution: smart grid technology requires creating new business models (2012). http://www.generatinginsights.com/whitepaper/evolution-smartgrid-technology-requires-creating-new-business-models

Combination of Standards to Support Flexibility Management in the Smart Grid, Challenges and Opportunities

Hisain Elshaafi[1]([✉]), Meritxell Vinyals[2], Michael Dibley[3], Ivan Grimaldi[4], and Mario Sisinni[5]

[1] Waterford Institute of Technology, Waterford, Ireland
helshaafi@tssg.org
[2] CEA-LIST, Gif-sur-yvette, France
[3] Cardiff University, Cardiff, UK
[4] Telecom Italia, Rome, Italy
[5] Smart Metering Systems Plc, Cardif, UK

Abstract. This paper presents the results of an assessment of a wide range of standards in the smart grid and telecommunications domains that may be jointly used to implement three use cases focusing on the use of the multi-agent systems paradigm in enhancing the smart grid particularly in the area of flexibility management. In addition to supporting a decentralised grid, multi-agent systems can improve other aspects such as reliability, performance and security. The paper identifies relevant standards based on a set of key smart grid use cases from the EU Mas2tering project. The evaluation aims to provide recommendations on combining those standards. This will enable new collaboration opportunities between grid operators, telecom and energy companies, both from technology and business perspectives. The set of telecommunication and energy standards is identified based on existing EU smart grid implementations, models (such as the Smart Grid Architecture Model) and reports published by the European and International standardisation bodies and coordination groups.

1 Introduction

Telecommunication and grid standards are important in addressing specific problems in smart grids. Harmonising those standards can help improve the implementation of smart grids and will allow the energy industry to evolve. In the context of Multi-Agent Systems (MAS), there has been an increased interest in this paradigm together with applications to many systems and domains such as diagnostics, monitoring, simulation, network control and automation [1] which rely on standards to deliver a foundation for market adoption and customer satisfaction. For companies, standards are essential for creating compliant, reliable and safe systems and products, which are compatible with existing systems and products. Regulators use standards to identify efficiency needs, to establish saving targets and to measure compliance. Researchers, on the other hand,

© ICST Institute for Computer Sciences, Social Informatics and Telecommunications Engineering 2017
J. Hu et al. (Eds.): SmartGIFT 2016, LNICST 175, pp. 143–151, 2017.
DOI: 10.1007/978-3-319-47729-9_15

are interested in standards to discover new and better techniques for technology development and deployment. Given the significant and growing role of the standardisation in the telecommunications and electricity network, it is critical that the main standardisation bodies and available standards are identified and examined. Substantial amount of research, development, and innovation activities have taken place in the smart grid landscape. A number of authorities have taken a leadership in the process of standards review, development and evolution. Mas2tering project [2] aims to develop a secure multi-agent holonic system for distributed management, control and optimization of the grid while providing an integrated platform for energy and telecom stakeholders [3]. The development of such platform requires understanding of the smart grid and telecommunication standards, their relationships and how they can be implemented.

This paper is structured as follows. Section 2 describes the use cases in Mas2tering project upon which the standards are evaluated and their relevance determined. Section 3 describes the approach taken in the standards' evaluation. Section 4 lists and categorises the standards that are evaluated. The evaluation and the findings are discussed in Sect. 5. Section 6 concludes the paper.

2   Use Cases

In order to understand the context of the work, this section provides a brief description of the Mas2tering objective and of the three use cases of the project. Mas2tering aims at developing tools and services to be implemented at LV grid level. The scope is to create local communities of end users represented by local aggregators to allow a more effective management of flexibility with respect to traditional centralised one. The home energy box and the MAS platform are the key technologies of the project, since their use enables the local optimization services. To assess the effectiveness of the solution, the project relies on three technical use cases (UCs). Each UC is an enabler to the subsequent one in order and addresses portions of the LV distribution network that is gradually wider:

UC1: Secure and effective connection of commercial home energy boxes with smart meter and consumption profile optimisation
This UC focuses on the Home Area Network (HAN) and the services that involve prosumers. It aims to demonstrate the interoperability between the HAN management system, the smart meter and a technical interface which allows the bi-directional communication between the prosumer and the rest of the LV grid. The communication is a prerequisite to the local optimization proposed in the other UCs and for the prosumer to enter the market of flexibility products.

UC2: Decentralised energy management in a local area with Multi-Agents
This UC focuses on the district, intended as a community of prosumers represented by a local aggregator. It aims to demonstrate that MAS optimisation performed at this local level is effective for energy management and local balancing, as an alternative to traditional centralised optimization. The objective is to maximise revenue for prosumers belonging to the local community when

coping with variable external conditions and conflicting requests from the Distribution System Operator (DSO) and other grid actors. This UC will involve the deployment and testing of software-controlled equipment connected to the local grid in testing laboratories supported by large European companies.

UC3: Enhancing grid reliability, performance and resilience

The UC can be considered as an extension of UC2 and tackles the LV grid, intended as the union of more local communities in a given area (represented by a MV/LV substation). The UC targets in particular DSOs and aims at demonstrating that the local optimization enabled in UC2, coupled with proper grid monitoring can be a cost-effective way to deal with local congestions and globally increase grid performances, reliability and resilience.

3 Standards Evaluation Methodology

In order to evaluate the relevance of the standards to Mas2tering project based on the three use cases and to establish the relevance of selected standards and technologies, an evaluation framework has been created as presented in Table 1.

Table 1. Standards evaluation framework

Criterion	Description
Adequacy	Suitability of the standard for the project use cases
Benefits	Advantages of using the standard and its strength
Problems	Disadvantages of using the standard and its possible shortcomings
Interoperability	The possibility of using the standard in conjunction with other telecom and/or energy standards
Recommendations	Suggestions regarding the approach towards using the standard in the project and relevant comments on the standard
Scope	Describes issues around the scope of the standard in relation to its applicability to the project and the extent to which it can be used
License	Access to the standard e.g. free or paid access, and where to find it
Usability	The level of the standard's ease of use and implementation
Security	The standard's security aspects

4 Standards Evaluated

This section lists several telecom and smart grid standards that are provided or recommended by standards bodies including International Electrotechnical Commission (IEC), the European Telecommunications Standards Institute (ETSI), the Institute of Electrical and Electronics Engineers (IEEE), the National Institute of Standards and Technology (NIST), The Union of the Electricity Industry- Eurelectric (EURELECTRIC), CEN-CENELEC-ETSI Smart Grid Coordination Group, and Smart Grid Interoperability Panel (SGIP). All the listed standards are evaluated as part of Mas2tering project and their relevance to the project use cases is detailed in Sect. 2. The standards are classified into categories for easier understanding of their area of application and usefulness. However, this categorisation does not mean that a standard falls exclusively under a specific category as overlaps exist in the scopes of the standards. The categories names are self descriptive. The categories and the evaluated standards are listed in Table 2. References to the standard sources are not included for brevity.

5   Evaluation of Standards

This section describes the evaluation of the standards and its findings.

5.1   Mapping to the OSI Model

In addition to the categorisation described above, the standards were mapped to the seven layers of the OSI model during the evaluation as shown in Fig. 1. The layered model helps analyse each of the standards. It also provides a global picture that assists in investigating how the standards and protocols will integrate together and affect each other's operation.

5.2   Discussion of Findings

This section discusses a number of standards that were found to be most relevant to the use cases with particular focus to multi-agent systems, integration, energy management and security. In general, some of the standards have wide scopes e.g. applicable to substations, domestic meters, while others have more specific technical features applicable to certain features of the Mas2tering platform.

Messaging and Communication. From the messaging perspective, FIPA-ACL [4] has support libraries covering the full network stack in the chosen MAS platform framework for Mas2tering, namely JADE (Java Agent Development Framework)[6], thus is a low overhead option. Add-ons are available to assist in the encoding and decoding of messages at the application layer and for security. However, no direct support for interfacing to smart grid devices would be achieved without further effort. Hence, a gateway would be needed if FIPA-ACL were to be used. The adoption of DLMS (Device Language Message specification) [7] in ACL content would facilitate mapping between protocols together with exchange transactions compliant to COSEM (COmpanion Specification for Energy Metering) seem pertinent. In the scope of a gateway, the use of the scalable and extensible framework ISO/IEC 15045 would seem desirable.

Table 2. Categories of standards evaluated

Standard category	Evaluated standards
Integration and interface	Foundation for Intelligent Physical Agents - Agent Communication Language (FIPA-ACL) [4]
	IEC 61968/61970/62325 - Common Information Model (CIM) [5]
	IEEE 1615 - Recommended Practice for Network Communications in Substations
	IEC 62541 - OPC Unified Architecture (UA)
	IEEE 2030 - IEEE Guide for Smart Grid Interoperability
Energy management	- Open Automated Demand Response Communications Specification
	Universal Smart Energy Framework (USEF)
	ISO/IEC 15067 - Home Electronic System (HES) Application Model
Smart metering	IEEE 1377 - Utility Industry Metering Communication Protocol Application Layer
	ISO/IEC 15045 - Home Electronic Systems Gateway
	ETSI TS 103 908 - Power Line Telecommunications (PLT) BPSK Narrow Band Power Line Channel for Smart Metering Applications
	ETSI TR 102 691 - Machine to Machine (M2M) Communications
	EN 13757 - Meter-Bus
	ETSI TR 103 240 - PLT for Smart Metering and Home Automation
	IEC 62056 - Electricity metering
	Smart metering equipment technical specifications
Smart grid monitoring and performance	IEC 60870 - Data Transmission Protocols for Supervisory Control and Data Acquisition (SCADA)
	IEEE 1250 - IEEE Guide for Identifying and Improving Voltage Quality in Power Systems
	IEEE 1159 - Recommended Practice for Monitoring Electrical Power Quality
	IEEE 1613 - Environmental and Testing Requirements for Communications Networking Devices Installed in Electric Power Substations
	IEEE P1547 - Conformance Test Procedures for Equipment Interconnecting Distributed Resources
	CLC TS 50549-1 - Requirements for generating plants to be connected in parallel with distribution
	IEEE 1646 - Communication Delivery Time Performance Requirements for Electric Power Substation Automation
	IEEE C37.1 - SCADA and Automation Systems
Networking	IEEE standard for Electric Power Systems Communications Distributed Network Protocol (DNP3)
	IEC 61850 - Communication Networks and Systems in Substations
	ZigBee
	Virtual Private Netowks (VPNs)
	IETF RFC 6272 - Internet Protocols for the Smart Grid
	ISO/IEC 14908 - Control Network Protocol
Physical and Data Link	Synchronous Optical Networking (SONET) and Synchronous Digital Hierarchy (SDH)
	Digital Subscriber Line (DSL)
	Power Line Communication (PLC) standards
	Mobile communication standards: GSM, GPRS, EDGE, LTE, WiMAX
	LANs: Ethernet, IEEE 802.11 (Wireless LANs)
	GS OSG 001 - Open Smart Grid Protocol (OSGP)
Security	IEC 62351 - Power systems management and associated information exchange Data and communications security
	NIST 7628 - Guidelines for Smart Grid Cybersecurity
	IEEE 1686 - Intelligent Electronic Devices Cyber Security Capabilities
	IEEE PC 37.240 - Cybersecurity Requirements for Substation Automation, Protection, and Control Systems
	ISO/IEC 27k series - Information Security Management Systems
	IEC 62443 - Industrial Network and System Security

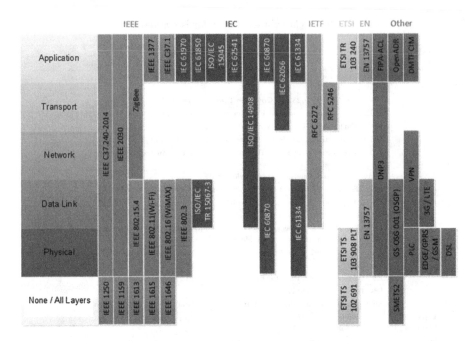

Fig. 1. Mapping of standards, protocols, and technologies to OSI model

The Common Information Model (CIM) [5] defined in IEC 61970 and supporting standards such as IEC 61968, is the main model on which FIPA-ACL message content will be based. CIM provides the basis for the design of generic communications for power systems, promoting interoperability. CIM describes elements of the electrical power system to support modelling to facilitate lifecycle stages such as operation of grids. It consists of three layers including the profile layer that allows the specification of sub models of the full specification to support information exchange for specific scenarios and contexts. It is expected that Mas2tering will specify one or more CIM profiles. The use of CIM thus simplifies the interoperability between different software applications delivering syntactic interoperability to support business procedures and objectives. Instead of defining mappings between every participant's internal representations involved in message exchanges and the necessary translations, the platform will map and translate to a common shared model. CIM which is widely reported in the context of the semantic smart grid, describes the smart grid domain but not its logical semantics, instead relying on implied semantics. Lack of semantic definitions may be adequate in some cases in Mas2tering and for that purpose a model with limited expressivity constructs can be used to avoid complicating the model in those contexts. However, where appropriate richer models can be formulated using a consistent representation but with the higher expressivity constructs.

IEC 61850 is an evolving standard with new object models that have been added or under development for electric components such as photovoltaic, batteries and electric vehicles. The use of XML and Ethernet in IEC 61850 contributes to an easier combination with the ICT asset. However IEC 61850 does not define demand response signals such as prices or curtailment signals. Some challenges exist to implementing IEC 61850 such as the extensiveness of the standard, the need for domain knowledge in the area of substation automation, ambiguity due to use of natural language description and inconsistency of different parts of the standard [8]. Although there is some overlap in the role of CIM and IEC 61850, they do not use the same languages. Efforts are underway to improve the interoperability between CIM and IEC 61850 [11].

ZigBee Home Automation (HA 1.2) [12] is a low power wireless sensor network protocol. It is designed for HAN devices, relying on IEEE 802.15.4 protocol specifications for MAC level, and a comprehensive ZigBee Cluster Library (ZCL). ZCL provides a list of functions already modelling and providing a lot of the required interactions between energy boxes and HAN devices such as the exchange of consumption profile and energy cost and the possibility to perform instantaneous demand. ZigBee HA will be used in UC1 to connect home devices with smart meter and the energy box. The main clusters specifying interactions between smart appliances and energy box are the ApplianceControl cluster and the PowerProfile cluster. ApplianceControl cluster enables the scheduling of smart appliances by exposing functions that allow the energy box to set the desired appliance cycle and its start time. The PowerProfile cluster allows the energy gateway to receive information from the appliances about their expected energy consumption in terms of peak power, expected duration and expected energy consumption for each phase of the scheduled cycle. Additionally, both smart meters and smart appliances can use the Metering cluster in order to provide information about the energy and power consumption.

OSGP protocol can be useful for addressing communication between smart meters and aggregators for pricing purposes. The standard includes interesting functionalities such as device discovery. However, there is a lot of room for improvement in this standard as it is missing elements required for the smart grid such as real time interaction with Distributed Energy Resources (DERs).

Flexibility Management. OpenADR aims to automate the demand response to control energy demand and it is among standards that were found to be very relevant to the Mas2tering project. OpenADR supports signals for electricity, energy prices, demand charges, customer bids, load dispatch, and storage. OpenADR "A" is sometimes called the bugged profile as the limited services or devices need the "B" profile to be operated in proper manners. We strongly recommend the developers to go for "B" profile and then narrow down the scale to profile "A" if the scenarios are not using some services. Moreover, profile "B" offers more security options than "A". The cybersecurity tasks will need to take in consideration the validation and the process of certificate provisioning on long periods. OpenADR is based on IP and uses either HTTP or XML. Although OpenADR has been approved as IEC/PAS 62746-10-1 by IEC in 2014, it is not considered as a core standard by the standardization body. There is currently work to adapt OpenADR to CIM that is a core standard [9]. This would make demand response compatible with utility standards.

Although not a standard, the USEF framework [10] is chosen to be used as a reference background to support the definition of the project's business framework and enable the comparison of Mas2tering project with other projects in the smart grid area. In particular, the market mechanisms described in USEF should be used as reference for the definition of the local optimization process. This would give concreteness to the Mas2tering solution from both technical and business perspectives, without imposing any constraint to its development.

Security. Security standards such as IEEE PC37.240, IEEE 2030 are assessed in relation to their usefulness in securing the Mas2tering platform. IEC 62351 can help secure the transactions between agents in the MAS environments. This standard is focused on implementing security for power system control operations. It is useful in securing the implementations for UC2 and UC3. IEEE PC37.240 and IEEE 2030 may be used as guidelines for the design of secure Mas2tering platform. The NIST 7628 guidelines can be used for securing UC1 e.g. HAN data exchange between devices and appliances.

6   Conclusion

This paper evaluates a number of important aspects of standards that are relevant to the smart grid. The aspects include adequacy to the use cases, problems, benefits, usability, interoperability and security. As a result of the evaluation, it has been decided that several standards are useful to the use cases including FIPA-ACL, IEC 61850, CIM, OpenADR, USEF, ZigBee, IEC 62351, IEEE PC37.240, IEEE 2030 and OSGP. This paper provides some of the recommendations regarding the combination and improvement of available telecommunication and energy standards such as IEEE PC37.240, IEC 62351 (for cybersecurity), CIM (for software and data model design), and IEEE 2030 standard (for interoperability). Importantly, gaps still exist in communication standards, protocols, and technologies particularly in relation to interoperability and security of communication systems and in harmonisation between models and standards despite of the significant work being carried out by the standardisation organisations worldwide.

Acknowledgement. The research leading to these results has partly received funding from the EU Seventh Framework Programme under grants n° 619682 (Mas2tering).

References

1. McArthur, S., Catterson, V., Davidson, E., Dimeas, A., Hatziargyriou, N., Ponci, F., Funabashi, T.: Multi-agent systems for power engineering applications. IEEE Trans. Power Syst. **22**, 1743–1752 (2007)
2. Multi-Agent Systems and Secured Coupling of Telecom and Energy Grids for Next Generation Smart Grid Services. http://www.mas2tering.eu
3. Mourshed, M., Robert, S., Ranalli, A., Messervey, T., Reforgiato, D., Contreau, R., Becue, A., Quinn, K., Rezgui, Y., Lennard, Z.: Smart grid futures: perspectives on the integration of energy and ICT services. Energy Procedia **75**, 1132–1137 (2015)
4. FIPA ACL Message Structure Specifications (2002). http://www.fipa.org/specs/ fipa00061
5. IEC Energy Management System Application Program Interface (EMS-API) - Part 301: Common Information Model (CIM) base (2013)
6. Bellifemine, F., Bergenti, F., Caire, G., Poggi, A.: JADE a Java Agent Development Framework. In: Multi-Agent Programming: Languages, Platforms and Applications. Number 15 in Multiagent Systems, Artificial Societies, and Simulated Organizations, chap. 5, pp. 125–148. Springer (2005)
7. DLMS user association. http://www.dlms.com
8. Y. Liang and R. Campbell, Understanding and Simulating the IEC 61850 Standard. https://www.ideals.illinois, 2008 edu/handle/2142/11457
9. CIM-OpenADR Harmonization. http://cimug.ucaiug.org/Projects/CIM-OpenADR
10. Smart Energy Collective. An Introduction to the Universal Smart Energy Framework (2014)
11. Santodomingo, R., Rodrguez-Mondjar, J.A., Sanz-Bobi, M.A.: Using semantic web resources to translate existing files between CIM and IEC 61850. IEEE Trans. Power Syst. **27**(4), 2047–2054 (2012)
12. ZigBee Alliance: ZigBee Home Automation Public Application Profile 1.2 (2013). http://www.zigbee.org

Special Session Track

A Load Balanced Charging Strategy for Electric Vehicle in Smart Grid

Qiang Tang[1](✉), Ming-zhong Xie[1], Li Wang[2], Yuan-sheng Luo[1],
and Kun Yang[3]

[1] School of Computer and Communication Engineering,
Changsha University of Science and Technology, Changsha, China
tangqiangcsust@163.com
[2] College of Computer Science and Technology, Taiyuan University of Technology,
Taiyuan, Shanxi, China
[3] School of Computer Science and Electronic Engineering,
University of Essex, Colchester, UK
kunyang@essex.ac.uk

Abstract. As the number of Electric Vehicle (EV) increases, the unco-
ordinated charging behaviors may cause the charging demand fluctuations and the charging load unbalanced. Besides, the users' charging
behaviors are affected by many factors. For example, the residual energy
of battery decides the travel distance of EV and if an EV has more
residual energy, the charging willing is lower. Because EV users don't
have much willing to change their charging time and place just as in the
past, the charging habit may also affect the charging decision. In this
paper, we propose a smart charging startegy CDF (Charging Decision
Function), where three sub-functions related to the residual energy of
battery, EV's charging habit, and the charging efficiency of charging station are all weighted and involved, for improving the balance of charging
load and reducing the charging demand fluctuations. The charging decision is resulted from the CDF's value, and if an EV decides to charge,
the charging time as well as charging place is also calculated. Compared
with other two related strategies, CDF has the best performance in terms
of reducing the charging demand fluctuations. The load balance among
different charging stations is also improved.

Keywords: Charging decision function · Load balance · Charging
demand fluctuations · Coordinated charging · Electric vehicle · Smart grid

1 Introduction

As the EV number increases, the power grid will suffer the overload and other passive effects [1]. For example a 20 % level of EV penetration would lead to a 35.8 %
increase in peak load [2]. Smart charging strategies have been studied in recent
years, and EVs are considered as important participants in terms of responding
to the electricity price i.e. demand response, voltage and participating in FR (frequency regulation) in Smart Grid [3]. Many strategies consider different indicators, such as the user's convenience is adopted in [4] and obtaining the minimum

© ICST Institute for Computer Sciences, Social Informatics and Telecommunications Engineering 2017
J. Hu et al. (Eds.): SmartGIFT 2016, LNICST 175, pp. 155–164, 2017.
DOI: 10.1007/978-3-319-47729-9_16

fluctuation load curve is the purpose in [5]. Besides the authors in [6] proposed the smart charging station to improve the efficiency of EV charging. In scheduling the EV charging load, the V2G [7], driving tours [8] and energy storage [9] can also be considered for improving the performance of charging strategy.

In recent years, many mathematical tools have been used to design the smart charging strategies. In [10], authors suggest the artificial intelligence technologies are efficient ways in designing the smart charging strategies. In general, the strategies are implemented as various control algorithms [11], for example the pricing algorithms are designed in [12] and in [13] the authors designed the stochastic systems with the relative high complexity. In order to reduce the complexity, the decentralized and distributed algorithms are designed [14].

Although, there is a lot of EV charging scheduling research work, the strategy about scheduling the EV charging load both in spatial dimension and time dimension is quite few. In this paper, we synthesize several aspects and propose an EV charging strategy CDF, which is a decision function containing the residual energy, charging habits and charging efficiency related weighted sub-functions, and based on CDF the EV users can easily decide where and when to charge and the balance of the charging load among charging stations are achieved.

2   Charging Decision Function based Electric Vehicle Charging Strategy

2.1   Charging Decision Function

The residual energy of battery, charging behavior, and charge station's efficiency are all combined into a CDF, which has three sub-functions: f_1, f_2 and f_3. In f_1, the residual energy of battery is involved. In f_2, the charging behavior is considered. In f_3, the charging efficiency is contained. The CDF is:

$$C(f_1, f_2, f_3) = \frac{2 \arctan(c_1 f_1 + c_2 f_2 + c_3 f_3)}{\pi}. \tag{1}$$

Where $C(f_1, f_2, f_3) \in (-1, +1)$, c_i is the weight of corresponding f_i. If we give a threshold θ, then a EV user is determined to charge while $C(f_1, f_2, f_3) > \theta$. In the next subsections, we will explain the three factors as well as sub-functions.

2.2   Residual Energy of Battery

We know that the travel distance of EV is uncertain. Let $p(x)$ denote the probability density function of the next trip's travel distance, and x is the distance of next trip, which is non-negative. Then we can get the probability distribution function:

$$P(m) = \int_0^m p(x)dx. \tag{2}$$

Where $P(m)$ represents a probability that the travel distance is less than m. In this paper, we assume the travel distance m is related to the residual energy of battery denoted by y, and the relationship can be expressed:

$$m = c_y y \pm \delta \qquad (3)$$

Where δ is the deviation and it satisfied $0 < \delta < \Delta$, and we assume the travel distance m is a uniform distribution, if the residual energy of an EV is y. In other words, for each distance m belonging to the interval $[c_y y - \Delta, c_y y + \Delta]$. For each distance $m_i \in [c_y y - \Delta, c_y y + \Delta]$, probability $P(m_i)$ can be calculated according to (2). Since m is a uniform distribution, then we get the probability density function for the travel distance to be

$$\varphi(m, y) = \begin{cases} \dfrac{1}{2\Delta}, & \text{if } m \in [c_y y - \Delta, c_y y + \Delta] \\ 0, & \text{else case} \end{cases} \qquad (4)$$

The excepted probability can be calculated according to:

$$EP(y) = \int_0^{+\infty} \varphi(m, y) dm \cdot \int_0^m p(x) dx. \qquad (5)$$

According to (5), if we give y a value, then the travel distance can be fixed into a range, and each value m_i of this range has a probability $P(m)$ according to (2). Because the travel distance is uniform distributed in this range, excepted probability can be simplified as

$$EP(y) = \frac{1}{2\Delta} \int_{c_y y - \Delta}^{c_y y + \Delta} dm \int_0^m p(x) dx. \qquad (6)$$

Because if y increases, the travel distance will increase, and $P(m)$ will increase, which cause the increasing of $EP(y)$. So, if the residual energy of battery is bigger, the excepted probability is smaller, and the charging decision value f_1 is also smaller. Then we set f_1 as:

$$f_1 = [1 - EP(y)]^n. \qquad (7)$$

Where n is a exponential, that way f_1 will be a nonlinear function.

In order to get the probability density function $p(x)$, we record the travel distances between two charging behaviors, which can indicate the travel distance for each charging. Assume the recorded travel distances as $x_j (j = 1, 2, \ldots, j_{max})$, and $p(x)$ is defined as:

$$p(x) = \omega_1 \sum_{j=0}^{j_{max}} \exp\{-\omega_2 (x - x_j)^2\}. \qquad (8)$$

Where ω_2 is a small constant, which is set in the simulation section. x is the travel distance of the next trip and belongs to $[0, M_{max}]$. According to (8) and properties of probability distributed function, we get:

$$\omega_1 = [\sum_{x=0}^{M_{max}} \sum_{j=0}^{j_{max}} \exp\{-\omega_2(x - x_j)^2\}]^{-1}. \tag{9}$$

Besides, in order to easily calculate (6), we set the travel distance precision as 1 km. Then (6) can be discretized as:

$$EP(y) = \frac{2}{M_{max}^2} \sum_{m=0}^{M_{max}} \sum_{x=0}^{m} \varphi(m, y) p(x) \tag{10}$$

2.3 EV's Charging Habit

We assume that the vehicle will not be charged during the traveling. We set three statuses for each EV, which are charging, driving and parking. For the charging habits, we only consider two factors, which are charging time t, and charging place $p = (p_x, p_y)$. In each specific time, the EV user gets a record $R = (t, b, p)$, and save it into a sequence. In the record, the charging behavior b has three values 1, 0 and -1, which stands for charging, parking and driving respectively.

The system learns the charging habits according to the records saved in the sequence. It at first learns in the time dimension:

$$C_t(t) = c_t \arctan \sum_{i=1}^{N} \frac{b^{[i]}}{(t - t^{[i]})^2}. \tag{11}$$

Where N is the number of training samples, and the latest N records will be selected as the training sample. c_t is a constant. t is the current time. We use arctan function to map the value into a controllable range $(-\frac{\pi}{2}, +\frac{\pi}{2})$.

In (11), the time cycle is a day, i.e. 24 h. We define that the time difference between one day's 23:00 and another day's 23:00 is zero. We also bound the time difference into 12 h:

$$\Delta t = \begin{cases} \Delta t, & \text{if } \Delta t < 12 \\ 24 - \Delta t, & \text{else case} \end{cases} \tag{12}$$

Where $\Delta t = |t - t^{[i]}|$. According to (12), the time difference between 23:00 and 2:00 is 3 rather than 21.

Similarly, the system also learns the charging habits in the space dimension:

$$C_p(p) = c_p \arctan \sum_{i=1}^{N} \frac{b^{[i]}}{\|p - p^{[i]}\|^2}. \tag{13}$$

Where c_p is a constant, p is the current position. The sub-function f_2 can be expressed:

$$f_2 = C_t(t) + C_p(p) - U. \tag{14}$$

Where U is a constant.

2.4   Charging Efficiency

In order to evaluate the charging efficiency, we define an efficiency index:

$$e = \begin{cases} \dfrac{\sigma}{\sum \sigma_i}, & \text{if } \sigma \leq \sum \sigma_i, \\ 1, & \text{else case.} \end{cases} \tag{15}$$

Where σ is the total power capacity of charging station and σ_i is the needed charging power of the ith EV which is connected with charging plot and waiting for charging at the charging station.

Besides, we assume there are L charging stations. For a specific EV, there are different distances between it and different charging stations. By considering both the distance and charging efficiency, we design the following charging decision sub-function:

$$f_3 = \alpha \left(\sum_{j=1}^{L} \frac{e_j}{\arctan d_j} \Big/ \sum_{i=1}^{L} \frac{1}{\arctan d_i} \right)^{\gamma} - \beta. \tag{16}$$

Where α, β and γ are parameters. According to (15), we know that if the charging efficiency e_j increases, the value of f_3 becomes larger. In order to express the distance influence on f_3, every charging station's e_j is divided by an arctan function with distance as an input parameter. In this paper, we set $\beta \geq \alpha$, which means the function f_3 is less than 0.

2.5   Making Charging Decision

If all the sub-functions are calculated, the charging decision function in (1) can be calculated also. By comparing the value of (1) and the value of θ, every EV user can decide whether to charge or not at the current time.

If one user decides to charge, then he/she should select a charging station. We propose the following selection function:

$$z_i = w_e e_i + w_d \frac{1}{\arctan d_i}, i \in [1, 2, \ldots, L]. \tag{17}$$

Where z_i is a weight of ith charging station and calculated by each EV. w_e is a weight of charging efficiency, and w_d is a weight of distance. Each EV will select the charging station with the maximum weight z.

3   Performance Evaluation

3.1   Environment and Parameters Settings

In this paper, we mainly simulate two types of EVs, which are private EVs and commercial EVs (or passenger vehicles) such as bus and taxi. The two types of vehicles have different driving pattern. The charging decision is made only when the user are controlling the car. As for the private cars, we assume the users are companies staffs, and have regular commuter time. For each private car user, we randomly generate the time point of going to work and the time point of going off work. A charging decision as well as the charging decision function is calculated only once in one hour. As for the commercial cars, we assume they travel in the whole day, and the charging decision function are calculated in every hour of the day. If an EV has decided to charge, it will select a charging station and go to the waiting zone of the charging station.

In order to meet the charging demand of these vehicles, we set 5 charging stations in the area, and each station can charge 200 vehicles and provides 200 waiting places. Every charging station adopts the alternating charging method. All the EV's charging power are the same.

When calculate the travel distance between the EV and charging station, we multiplied the straight-line distance by a random multiplier uniform distributed in $[1.4, 1.7]$. The average energy consumption is set to $5.05\,km/(kwh)$ according to the battery capacities and driving ranges in [15]. SOC (State Of Charge, which is a ratio between the residual energy and the total capacity of battery and belongs to the interval $(0,1)$.) is set to a uniformly distribution in $[0.3, 0.7]$. The area is square area with the range as $15\,km \times 15\,km$. With the negative exponential distribution, each private EV randomly generates a home coordinate and a company coordinate. The battery capacities are set as: $40\,kwh$, $44\,kwh$, $47\,kwh$, $53\,kwh$. All the simulated EVs are randomly assigned a capacity. The other needed parameters are set in the following Table 1:

Table 1. Parameter settings

PARM	c_1	c_2	c_3	n	γ	c_t	c_p	U
VALUE	1	0.2	0.7	5	5	0.5	0.5	1
PARM	c_y	Δ	ω_2	M_{max}	ω_e/ω_d	α	β	θ
VALUE	5.05	0.35	10^{-6}	300km	3	1	1	0.1

3.2 Compared Strategies

In order to compared with other strategies, we define a basic charging strategy (BCS). In the BCS, an EV will decide to charge only if its SOC is lower than the lower limit and select the nearest charging station to charging. There is no coordination between the charging station and the EV.

Another strategy called as random access strategy (RAS), which is a distributed algorithm, and by designing the probability function, each EV can be independently charged in accordance with the probability. In order to be suit to the scenario, the access probability function is:

$$P_A(SOC) = \exp\{-1.5 \cdot SOC\} \tag{18}$$

Where SOC is set as a uniformly distributed number in the interval $[0.3, 0.7]$. If an EV has calculated out the P_A, it compares it with a randomly generated number r_1 and if $P_A > r_1$, the variable C_s is set as 1.

In order to delay charge at the charging station with low charging efficiency, we set the another probability function related with charging efficiency:

$$P_E(W_E) = \exp\{-0.6 \cdot W_E\} \tag{19}$$

Where W_E is a weighted charging efficiency which can be calculated by:

$$W_E = \sum_{j=1}^{L} \frac{e_j}{\arctan d_j} \Big/ \sum_{i=1}^{L} \frac{1}{\arctan d_i}. \tag{20}$$

If an EV has calculated out the P_E, it compares it with a randomly generated number r_2, and if $P_E < r_2$, the variable C_e is set as 1. The final charging decision C_d is:

$$C_d = C_s \& C_e. \tag{21}$$

If C_d is 1, the EV decides to charge. Table 2 shows the comparison:

Table 2. The comparison of EV charging strategies

	BCS	RAS	CDF
Residual energy	Yes	Yes	Yes
Charging efficiency	No	Yes	Yes
Charging habits	No	No	Yes

3.3 Charging Demands Fluctuations

We designed a scenario which contains 2500 private EVs and 2000 commercial EVs. We simulate 72 h, and in every hour the two types EVs' total charging demand as well as the total charging demand are calculated once. Accordingly we get the 72 charging demand points for each charging strategy. Simulation results are shown in Fig. 1.

According to the Fig. 1(a), we find that the BCS strategy can not shift the charging demand efficiently, because it only consider the SOC statues, and other factors such as charging efficiency are all not considered at all. Figure 1(b) shows that the RAS strategy has a certain ability to shift the charging demand, which is because RAS considers both the SOC and charging efficiency of charging station. If the private EVs have a charging peak, the RAS schedules the commercial vehicles to charge at another time slot. Because the CDF has considered the residual energy, charging efficiency as well as the charging habits, and all of which are weighted and combined into a decision function, thus it has the strongest ability to shift the charging demands, which is shown in Fig. 1(c) and the peak of private car's demand curve appears at the time when the valley of commercial vehicle's demand curve appears. Figure 1(d) shows that the CDF has the minimum charging demands fluctuations, which is because the CDF has the strongest ability to shift the charging demands and the demands curve is more flat compared with other two strategies.

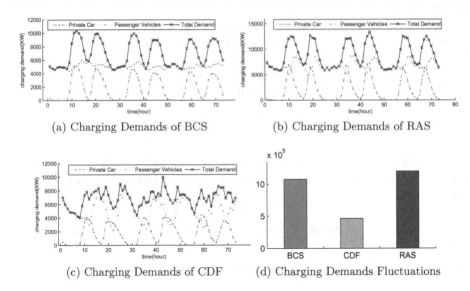

(a) Charging Demands of BCS   (b) Charging Demands of RAS

(c) Charging Demands of CDF   (d) Charging Demands Fluctuations

Fig. 1. The charging demands of different strategies

3.4 Load Balance

In order to introduce the load balance among different charging stations, we calculate the charging loads of different EV types for each charging station. The total charging demands for each charging station is also presented. Results are shown in Fig. 2:

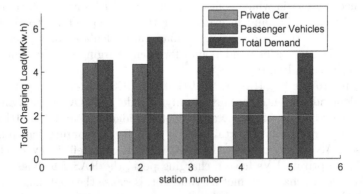

Fig. 2. The load balance among charging stations.

According to the Fig. 2, we find that the total demands of different charging stations are almost balanced, which is because the CDF has consider the charging efficiency and if the charging station has less charging loads, the charging efficiency is larger, which results in attracting more EVs to select this charging station to charge.

4   Conclusion

In this paper, we design a distributed charging strategies CDF for electric vehicle, which is supported by a Charging Decision Function. By considering the residual energy of the battery, charging habits, charging efficiency, CDF has strong ability to shift the charging demands and reduce the fluctuations. In the future work, we will continue focus on the different charging prices of different charging stations, and the travel speed is also another important factor will be considered further.

References

1. Gong, X., Lin, T., Su, B.: Survey on the impact of electric vehicles on power distribution grid. In: Proceedings of the Power Engineering, Automation Conference (PEAM), pp. 553–557. IEEE (2011)
2. Qian, K., Zhou, C., Allan, M., et al.: Modeling of load demand due to EV battery charging in distribution systems. IEEE Trans. Power Syst. $26(2)$, 802–810 (2011)
3. Jin, Z., Lina, H., Canbing, L., et al.: Coordinated control for large-scale EV charging facilities, energy storage devices participating in frequency regulation. Appl. Energy $123(15)$, 253–262 (2014)
4. Karfopoulos, E.L., Hatziargyriou, N.: Distributed coordination of electric vehicles providing V2G services. IEEE Trans. Power Syst. $31(1)$, 1–10 (2016)
5. Gharbaoui, M., Bruno, R., Martini, B., et al.: Assessing the effect of introducing adaptive charging stations in public EV charging infrastructures. In: Proceedings of the 2014 International Conference on Connected Vehicles, Expo (ICCVE), pp. 299–305 (2014)
6. Binetti, G., Davoudi, A., Naso, D., et al.: Scalable real-time electric vehicles charging with discrete charging rates. IEEE Trans. Smart Grid $6(5)$, 2211–2220 (2015)
7. Francesco, M., Claudio, C., Carla-Fabiana, C., Massimo, R.: A game-theory analysis of charging stations selection by EV drivers. Int. J. Perform. Eval. $83–84$, 16–31 (2015)
8. Joana, C., Goncalo, H., Joao, G.: A MIP model for locating slow-charging stations for electric vehicles in urban areas accounting for driver tours. Transp. Res. Part E: Logistics Transp. Rev. ScienceDirect 75, 188–201 (2015)
9. Sbordone, D., Bertini, I., Di Pietra, B., et al.: Fast charging stations, energy storage technologies: A real implementation in the smart micro grid paradigm. Electr. Power Syst. Res. 120, 96–108 (2015)
10. Liu, Y., Li, Q., Tao, S., et al.: Coordinated EV charging, its application in distribution networks. In: Proceedings of the 2013 International Conference on Technological Advances in Electrical, Electronics, Computer Engineering (TAEECE), pp. 600–604 (2013)
11. Rigas, E.S., Ramchurn, S.D., Bassiliades, N.: Managing electric vehicles in the smart grid using artificial intelligence: a survey. IEEE Trans. Intell. Transp. Syst. $1(4)$, 1619–1635 (2015)
12. Cao, Y., Tang, S., Li, C., et al.: An optimized EV charging model considering TOU price, SOC curve. IEEE Trans. Smart Grid $3(1)$, 388–393 (2012)
13. Jaeyoung, J., Chow, J.Y.J., Jayakrishnan, R., et al.: Stochastic dynamic itinerary interception refueling location problem with queue delay for electric taxi charging stations. Transp. Res. Part C: Emerg. Technol. 40, 123–142 (2014)

14. Zhou, K., Cai, L.: Randomized PHEV charging under distribution grid constraints. Trans. Smart Grid **5**(2), 879–887 (2014)
15. Frieske, B., Kloetzke, M., Mauser, F.: Trends in vehicle concept, key technology development for hybrid, battery electric vehicles. In: Proceedings of the 2013 World Electric Vehicle Symposium, Exhibition (EVS27), pp. 1–12 (2013)

Optimized Energy-Aware Window Control for IEEE 802.11ah Based Networks in Smart Grid Enabled Cities

Yanru Wang[1(✉)], Chao Liu[2], Kok Keong Chai[1], Yue Chen[1], and Jonathan Loo[3]

[1] Queen Mary University of London, London E1 4NS, UK
{yanru.wang,michael.chai,yue.chen}@qmul.ac.uk
[2] International School, Beijing University of Posts and Telecommunications, Beijing, China
liuchao@bupt.edu.cn
[3] Middlesex University, London NW4 4BT, UK
J.Loo@mdx.ac.uk

Abstract. IEEE 802.11ah brings Restricted Access Window (RAW) to decrease collision probability for smart grid applications. The RAW size affects transmission energy consumption and data rate for the different number of devices per group. In this paper, we investigate an energy efficient RAW optimization problem for IEEE 802.11ah based uplink communications. We formulate the problem based on overall energy consumption and the data rate of each RAW by applying probability theory. Then, we derive the energy efficiency of the uplink transmission. Last but not the least, a dynamic energy-aware window algorithm to adapt the RAW size is proposed to optimize the energy efficiency by identifying the number of slots in each RAW for different group scales. Simulation results show that our proposed algorithm outperforms existing RAW on uplink energy efficiency and delivery ratio.

Keywords: Smart grid · IEEE 802.11ah · RAW · Uplink energy efficiency

1 Introduction

Smart grid provides intelligent transfer and usage of power and energy in cities, indicating there will be a large number of sensors and devices [1,2]. So robust information and communication technologies play a significant role for constructing such networks [3]. However, existing wireless technologies, such as RFID, ZigBee, Bluetooth, etc., can not accommodate such a high density of devices with high throughput over a large transmission range [4].

One of the latest wireless communication technologies that been proposed for smart grid application is the Low Power Wi-Fi [2], as shown in Fig. 1. IEEE 802.11ah Wireless LAN standard group has put forward IEEE 802.11ah to support Low Power Wi-Fi, which started in November 2010 and is expected to finish

© ICST Institute for Computer Sciences, Social Informatics and Telecommunications Engineering 2017
J. Hu et al. (Eds.): SmartGIFT 2016, LNICST 175, pp. 165–173, 2017.
DOI: 10.1007/978-3-319-47729-9_17

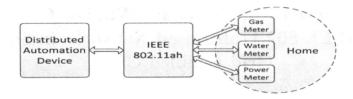

Fig. 1. Smart grid infrastructure applied IEEE 802.11ah.

not before 2016 [5]. IEEE 802.11ah operates at sub-1 GHz and it can support up to 6000 devices within a network with transmission range up to 1 Km at the rate of more than 100 kbps [6]. Restricted Access Window (RAW) is a new MAC layer feature that used in Low Power Wi-Fi to decrease collision. It limits a set of devices that can access the channel at any time and spreads their attempts over a long period of time [5,7]. RAW consists of multiple equal time slots, where each slot is selected by devices or assigned to a group of devices for transmission [4,8]. Devices are in wake-up mode only when turning to their RAW, otherwise would be in doze mode.

The RAW size has not been defined in the standard, which has the influence on energy efficiency. With long duration RAW, the devices involved should be in active mode for a longer period, leading to expending idle wake-up energy. On the contrary, the collision probability would be high if a large number of devices access through RAW with limited time slots, resulting in low efficiency. RAW also affects on overhead information to inform scheduling information. Thus adaptive RAW duration for diverse number of devices per group could reduce low energy consumption and achieve a high data rate.

The way to improve energy efficiency has been studied in depth in many research works, mainly focuses on improving successful transmission probability and reducing collision. In [8], a new medium access control enhancement algorithm was proposed to find optimal size of RAW by Maximum Likelihood (ML) estimation method. However, it only has not involved consideration of energy. In [9], the authors introduced Successive Interference Cancellation to improve the throughput in limited time but this could result in more collisions. In [10,11], new algorithms were proposed to calculate wake-up time of devices by probability theory and matrix way to build analysis model. However, these algorithms have not embodied the RAW communication mechanism. In [12], low collision probability was achieved through access control to limit the number of devices contending in authentication stage. However, it was not suitable for transmitting and receiving process.

The aforementioned literatures laid a solid foundation in improving energy efficiency based on transmission probability and reducing collision for IEEE 802.11ah. Less work has been done to optimise the energy efficiency by dynamic RAW size based on clustering size. To address the above joint consideration, in this paper we study an optimisation problem aiming at maximizing uplink energy efficiency through RAW. A dynamic energy-aware window algorithm is proposed to determine the RAW duration for uplink communications for different group sizes. An optimal solution is derived by Gradient Descent, a fast method with less complexity to solve optimization problem.

2   System Model

We consider a single-hop topology for dense IEEE 802.11ah smart grid networks as a single Access Point (AP) with a high number of devices. RAW groups the devices and splits the channel into equal time slots [13]. All devices in this network listen in the beginning of beacon frame called Target Beacon Transmission Time (TBTT) to obtain scheduling information that indicates which RAW they belongs to. Then devices would fall into sleep mode until turning to their RAW to attempt accessing. For each RAW, there are M time slots and N devices limited by lowest and highest Associated Identifier (AID) of devices which indicate the location, traffic, type, energy saving mode etc. [4].

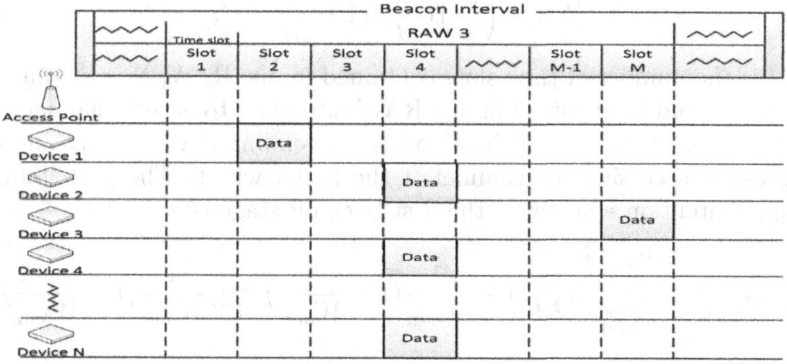

Fig. 2. Operation of RAW.

In uplink communications, the devices who have buffered data for the AP select a time slot of their RAW randomly and attempt to access channel as shown in Fig. 2. If there is only one device in a slot, it could access the slot directly, for example, Device 1 transmits packet in Slot 2 directly without contention, and the same for Devices 3 in Slot M. When there are more than one device choosing the same time slot, for example Device 2, 4 and N in Slot 4, they would go into the back-off stage to avoid collision by doubling contention window and trying again until reaching the slot boundary. If accessing successfully, device requests uplink communication by sending PowerSave-poll (PS-poll) message to the AP. AP responses with an ACK to confirm connection. After the first handshake, the device transmits buffered data frame and waits for ACK from AP [8]. The process is repeated, one RAW by one RAW, until the end of beacon frame.

3 Problem Formulation

As for uplink communication in MAC layer of IEEE 802.11ah, time slots in RAW are selected by devices randomly. According to [8], there are two cases for a device to transmit uplink packet successfully.

Case 1: a time slot chosen by only one device. This device would transmit a packet successfully without other contending devices.

Case 2: a time slot chosen by multiple devices. The devices will go into back-off stage, and one of them will succeeds in accessing at the first back-off stage.

For a single device, it could choose any time slot within one RAW. The probability of a time chosen by only one device is (Case 1):

$$P_1 = (1 - \frac{1}{M})^{N-1}. \tag{1}$$

By random selection, the probability of a time slot chosen by a number of devices, i is:

$$P(i) = \binom{N}{i}(\frac{1}{M})^i(1 - \frac{1}{M})^{N-i}, \tag{2}$$

where M is the number of time slots contained in one RAW; N is the number of devices that could be involved in one RAW intending to access channel.

For the Case 2, in view of (i−1) other contending devices, only one device will success in accessing the channel at the first back-off. The probability of a minimum contention window as the first back-off stage is:

$$P_{back-off}(i) = \sum_{k=0}^{W_{min}-1} \{\prod_0^k [1 - \frac{1}{W_{min}}(1 - \frac{k}{W_{min}})^{i-1}]\}\frac{1}{W_{min}}(1 - \frac{k+1}{W_{min}})^{i-1}, \tag{3}$$

where W_{min} is the minimal size of contention window.

So when there are i devices selecting the same time slot in one RAW, a device accessing channel successfully is based on the probability as $P_2(i) = P(i)P_{back-off}(i)$, i is from 2 to N.

Thus the overall probability for Case 2 is

$$P_{2\_all} = \sum_{i=2}^{N} P_2(i). \tag{4}$$

The successful transmission probability for one device to transmit one packet is the sum of two cases, which could be denoted by $P = P_1 + P_{2\_all}$.

Based on different states a device may fall into, the energy consumption of one device to transmit single packet in one RAW is

$$E = P_1 E_{s1} + P_{2\_all} E_{s2} + (1 - P)E_c + M E_{con}, \tag{5}$$

where E_{s1} is the energy consumption when transmitting a packet as Case 1; E_{s2} is the amount of energy consumed if transmitting as Case 2; E_c is the energy waste when there is collision so that it needs to retransmit in another RAW; E_{con} is the contention power, which is the energy consumed in wake-up mode.

The RAW size also determines the energy consumption of transmitting overhead information. For a short window duration, the overhead information of each device would be high due to the scheduling information that needs to be transmitted multiple times in a short time. And if the number of devices involved in one RAW is small, it also needs massive scheduling information to realize network communication. So the energy consumption of header information is related to N and M:

$$E_{head} = \frac{\alpha}{M} \times \frac{\beta}{N}, \tag{6}$$

α is the parameter indicating traffic and β is the parameter related to overall number of devices in the scenario.

Energy efficiency of one RAW could be evaluated by the data rate it provides and overall energy consumption. Data rate could be formulated as

$$R = \frac{N \times P \times \gamma}{\tau M}, \tag{7}$$

where γ is the packet size and τ is the time duration of one time slot. $N \times P \times \gamma$ is the total length of packets could be transmitted for N devices in M time slots. τM is the total time of one RAW.

The overall energy consumption consists of transition power and overhead power when N devices attempt to communicate with AP during one RAW, which could be denoted by

$$E_{overall} = N(E + E_{head}). \tag{8}$$

Thus energy efficiency is

$$EE(M) = \frac{R}{E_{overall}} = \frac{\gamma P}{\tau M(E + E_{head})}. \tag{9}$$

With P, E and E_{head} being built by M and N, energy efficiency is a function related to the number of devices involved and time slots in one RAW. We could maximize energy efficiency by finding optimal M based on N.

4   Dynamic Energy-Aware RAW

In this section, optimisation with RAW duration based on the number of devices per group is presented.

The main part of energy efficiency $f(M)$ could be denoted by

$$f(M) = \frac{P}{M[E + E_{head}]}. \tag{10}$$

Due to P_1 and P_{2_all} could be regarded as binomial distribution, if N is large (i.e., $N \geq 20$), the expression can be approximated by the Poisson distribution:

$$P_1 = Ne^{-\frac{N}{M}}, \tag{11}$$

$$P_{2\_all} = \sum_{2}^{N} P_{back\_off}(i) \frac{\left(\frac{N}{M}\right)^k e^{-\frac{N}{M}}}{k!}. \tag{12}$$

So the $f(M)$ can be simplified as

$$f(M) = \frac{Ne^{-\frac{N}{M}}}{(M)^2 E_{con} + MNE_{s1}e^{-\frac{N}{M}} + ME_c} \tag{13}$$

The second order derivation is the main factor to show function's concavity and convexity. Through operation, the result of second order derivation could be simplified as

$$f''(M) = \frac{e^{\frac{-N}{M}} N^2 \ln e(Nlne - 2M)}{M^4 [E_c M + E_{s1} e^{\frac{-N}{M}} + \tau E_{con}(M)^2]}, \tag{14}$$

which is negative, indicating it is a concave curve along different RAW duration with one peak point.

We find optimal RAW size for different number of devices per group to maximize energy efficiency by applying Gradient Descent as shown in Algorithm 1.

Algorithm 1. Gradient Descent Dynamic Energy-aware Window Algorithm

---

1: **Access point identify the number of devices per group N.**
2: **loop**
3:     **Find the optimal RAW size M for N devices per group based on energy efficiency.**
4:     Initialize M_{old} and M_{new} as two random numbers.
5:     $EE\_derivative(M) = Diff\{-EE(M), M\}$
6:     **while** $abs\{EE(M_{old}) - EE(M_{new})\} \geq precision$ **do**
7:         $\partial = 0.01$
8:         $M_{old} = M_{new}$
9:         $M_{new} = M_{old} - \partial \times EE\_derivative(M)$
10:    **end while**
11:    return M
12:    N devices randomly select time slots in RAW with M time slots and attempt to do uplink communication with AP.
13: **end loop**

This algorithm is a Gradient Descent approach to find the optimal solution, which is a method with less working and storage space. It begins with an initial value and finds the optimal value according to the gradient descent route.

5 Simulation Result and Analysis

In this section, the optimized RAW control algorithm is evaluated in Matlab. We consider a one-hop topology of one AP with multiple devices as describe

Table 1. Simulation parameters

Parameter	Value	Parameter	Value	Parameter	Value
Frequency	0.9 GHz	Data rate	100 kbps	Transmit power	1.346 mw
Transmit power in back-off stage	2.5 mw	Collision power	3.0 mw	Idle listen power	0.001 mw
Min contention window	8	Max contention window	1024	Packet length	1024 bits
Slot duration	31.1 ms	α	200	β	200

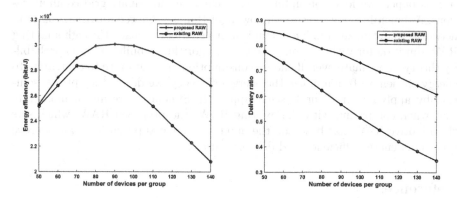

Fig. 3. Energy efficiency comparison with existing RAW.

in the system model. We assume every device involved has exactly one packet for uplink communications during a RAW. The main simulation parameters are given in Table 1.

The energy efficiency and delivery ratio varying over the diverse number of devices per group are shown in Fig. 3. The number of time slots in existing RAW is fixed [12], while the proposed one sets the RAW size based on the group scale. For energy efficiency, the proposed RAW outperforms the existing one with the improvement of 20% in general. The peak point of existing RAW is at $N = 70$, while that of proposed RAW is at $N = 90$ and 7% higher. When the number of devices is low, the trends of energy efficiency for both two RAW control go up since with an increase in the number of devices, the overhead information do not need to be sent multiple times to realize communications for whole networks, which lowers the energy consumption in informing scheduling. After the peak, more devices per group would lead to high collision probability, which results in low energy efficiency, so the trends decline. And the proposed RAW could set the RAW dynamically based on different scenarios, thus more time slots would be set in one RAW for lager group scale, bringing in higher energy efficiency.

As for uplink packet delivery ratio in one RAW, the proposed one could improve in general 50% when comparing with the existing RAW. With the rising number of devices per group, the trends of two curves go down due to higher collision probability which leads to consume more energy and less data that be successfully transmitted. The number of time slots in proposed RAW would be added for more devices involved to reduce contention, while the size of existing one is fixed no matter how many devices per group, so the rate of decrement in proposed one is lower than the existing one and the improvement is larger for higher density of devices per group.

6   Conclusions

In this paper, we focus on uplink energy efficiency for smart grid communications. A dynamic energy-aware window algorithm is proposed for IEEE 802.11ah networks to optimize uplink communications energy efficiency through adapting RAW duration for various group scale. The algorithm is built based on probability theory to estimate overall energy consumption and data rate to contribute to energy efficiency. To maximize the energy efficiency, we derive the optimal solution by applying Gradient Descent approach. Simulation results demonstrate that when comparing with the existing RAW, the proposed RAW, which sets the adaptive RAW size based on the number of devices per group, can improve the uplink energy efficiency and delivery ratio.

References

1. Sum, C.-S., Harada, H., Kojima, F., Lan, Z., Funada, R.: Smart utility networks in TV white space. IEEE Commun. Mag. **49**, 132–139 (2011)
2. Aust, S., Prasad, R., Niemegeers, I.G.M.M.: IEEE 802.11ah: advantages in standards and further challenges for sub 1GHz Wi-Fi. In: 2012 IEEE International Conference on Communications (ICC), pp. 6885–6889 (2012)
3. Habash, R.W.Y., Groza, V., Krewski, D., Paoli, G.: A risk assessment framework for the smart grid. In: 2013 IEEE Electrical Power and Energy Conference (EPEC), pp. 1–6 (2013)
4. Khorov, E., Lyakhov, A., Khorov, A., Guschin, A.: A survey on IEEE 802.11ah: an enabling networking technology for smart cities. Comput. Commun. (2014)
5. IEEE Standard for Information Technology Telecommunications and information exchange between systems Local and metropolitan area networks Specific requirements Part 11: Wireless LAN Medium Access Control (MAC) and Physical Layer (PHY) specifications Amendment 10: Mesh Networking (2011)
6. Hazmi, A., Rinne, J., Valkama, M.: Feasibility study of IEEE 802.11ah radio technology for IoT and M2M use cases. In: 2012 IEEE Globecom Workshops (GC Wkshps), pp. 1687–1692. IEEE (2012)
7. Seok, Y.: Backoff procedure in RAW (2013). http://mentor.ieee.org/802.11/dcn/13/11-13-0080-00-00ah-backoff-procedure-in-raw.ppt
8. Park, C.W., Hwang, D., Lee, T.: Enhancement of IEEE 802.11ah MAC for M2M communications. Commun. Lett. **18**, 1151–1154 (2014)
9. Vazquez-Gallego, F., Rietti, M., Bas, J., Alonso-Zarate, J., Alonso, L.: Performance evaluation of frame slotted-ALOHA with successive interference cancellation in machine-to-machine networks. In: European Wireless, pp. 403–408 (2014)

10. Liu, R.P., Sutton, G.J., Collings, I.B.: Power save with offset listen interval for IEEE 802.11ah smart grid communications. In: 2013 IEEE International Conference on Communications (ICC), pp. 4488–4492. IEEE (2013)

11. Liu, R.P., Sutton, G.J., Collings, I.B.: WLAN power save with offset listen interval for machine to machine communications. Wirel. Commun. **13**(5), 2552–2562 (2014)

12. Zhou, Y., Wang, H., Zheng, S., Lei, Z.Z.: Advances in IEEE 802.11ah standardization for machine-type communications in sub-1GHz WLAN. In: 2013 IEEE International Conference on Communications Workshops (ICC), pp. 1269–1273. IEEE (2013)

13. Kim, J.: RAW assignment follow up (2013). http://mentor.ieee.org/802.11/dcn/13/11-13-0510-01-00ah-raw-assignment-follow-up.ppt

Smart Home System Network Architecture

Chenqi Yang[✉], Emilio Mistretta, Sara Chaychian, and Johann Siau

College Lane, Hatfield, Hertfordshire AL10 9AB, UK
c.yang21@herts.ac.uk

Abstract. A four-tier based Smart Home System is proposed in this paper; this type of network architecture is inspired by both the concept of the Internet of Things (IoT) and Machine to Machine (M2M) technologies. The main purpose of this structure is to establish a centralized resource management system, which can monitor and control each home appliance within a house. This aim and structure can help to overcome the most notable research challenge which is called interoperability of the system. The interoperability can be reflected from how other services such as assisted living and vehicle tracking are able to share and utilize the same resources. One of the structure advantages is to help existing house owners to live in a modern and convenient environment; it is also environmentally friendly by reducing the overall energy consumption within a household. A novel Home System Data Reflect Arc is proposed to minimalize the response time and limit unnecessary data transmission. Finally, with the aid and optimization of wireless sensor networks, a Received Signal Strength Indicator (RSSI) based indoor localization approach is proposed which is novel based on the fact that it does not require specialized implementation, nor does it require people to wear specialized hardware as it is part of a passive device free environment.

Keywords: Smart home automation · Indoor localization · RESTful web server · Cloud computing · Internet of Things (IoT) · Machine to Machine (M2M) · Received Signal Strength Indicator (RSSI) · Smart Home System (SHS)

1 Introduction

The main technologies discussed in this paper are a combination of local wireless sensor networks and public cloud computing platforms. The intelligent aspect of this structure is reflected by how information is gathered, transmitted and processed within the system and also how the system reacts to its ever changing surroundings. More importantly is the system ability to accurately predict individual's demands. The Smart Home System (SHS) can have a positive influence on energy efficiency and become environmentally friendly; this is possible by using energy only when it is required. One way of doing this is to reduce the operating hours of home appliances, in addition, SHS also provides a platform that allows other smart systems to interact with each other; such systems could include health care or vehicle monitoring. The more information and data that a SHS can obtain, the greater the chance of optimizing and improving the daily lives of individuals, there are many factors that can influence this.

© ICST Institute for Computer Sciences, Social Informatics and Telecommunications Engineering 2017
J. Hu et al. (Eds.): SmartGIFT 2016, LNICST 175, pp. 174–183, 2017.
DOI: 10.1007/978-3-319-47729-9_18

SHS is designed to establish a seamless data communication among home appliances within a local home domain, as well as a resource sharing based cloud computing platform. This communication channel allows each individual devices or control systems to be aware of each other's presence, so that related operations can be carried out by the system without necessary human interaction. On the other hand, a cloud based service offers SHS a more secure and restricted data exchange architecture; however, this standard public protocol brings many application scenarios and makes the task of third-party integration much easier, as detailed hardware level control instructions are now hidden from the user.

The first half of this paper focused on the Smart Home System (SHS) architecture and how different system layers are linked. The second half (Sect. 3) will focus on the sensor layer of the structure which is based on wireless sensor networks.

2   Smart Home System Architecture

From a global point of view, the proposed SHS can be divided into two functional areas: cloud computing (server) and home control (client). There is only one cloud implementation entity which consists of various types of server and databases, these are mainly responsible for data storage, information sharing and processing. However different implementations of home control systems can be deployed based on each home environment. The Internet infrastructure is also used as the physical link between the cloud platform and individual home control systems, so that IP communications can be made between the server and the client.

From a user's perspective, the primary focus is on the surrounding environment and the final outcomes that SHS can deliver, however from a developer's point of view this can be divided into four parts for the purpose of low coupling. The proposed structure is based on a multi-tier architecture, which includes data access, cloud logic, home logic and a sensor layer. Figure 1 illustrate an example of this structure and a general description is presented below:

- Data Access Layer: dedicated to data storage
- Cloud Logic Layer: information sharing and processing
- Home Logic Layer: links between human and sensor devices
- Sensor Layer: environment control and monitor

Using four-tier structure instead of three-tier architecture is the most notable feature of the system which makes it different from the other researches. A four tier approach can better deal with system data traffic a faster response time [1–4]. A three-tier structure is more suitable for web based technology where information is first sent to the server, processed and then the result displayed. However, the SHS is more like a distributed network, where each component within the system needs to communicate and coordinate its actions in order to achieve a common goal. In addition, SHS is not a centralized control system and the cloud platform has absolutely no control over the local household due to the complexity of the home environment and its different preferences. This makes the SHS impractical to produce universal control logic, furthermore integrating cloud communication, appliance control and sensor monitoring will increase the structure dependency on various parts of the system therefore making maintenance more difficult.

A main characteristic of the SHS architecture is loose coupling; by reducing the risk of changes made in one part of the system, errors within other parts can be prevented. Instead of deploying one business logic layer as a whole, two logic layers are located on the cloud and the local home environment, a separate sensor layer has been added also. The two logic layers are equally important and there is no affiliation between them, action synchronization is achieved by data communication through an IP network. Since the home control is able to operate independently, this independence also means that internet connectivity is not always required.

Scalability allows the system to handle the increasing amount of required functionalities in a capable manner. To scale horizontally (e.g. integrating similar functionalities into one layer) multiple home control units can be deployed into a wide range of areas in order to promote the smart community concept. However due to the number of deployments it is impossible to consider all equipment and add any new features to the existing home logic. In contrast a cloud server is a much smaller domain to work with; also its faster physical hardware and optimized software can be utilized to handle the ever increasing data traffic. Vertical scalability which makes use of and expands current infrastructure is essential when SHS needs to cooperate with other systems; (since external information needs to be translated into a format that SHS can understand, a new logic layer is needed to perform this procedure). As there is no direct dependence, layers only exchange information through a predefined communication channel, therefore new logic can be deployed without affecting other functional areas.

2.1 Reflex Arc in SHS

Advantages mentioned above are inherited from the multi-tier structure, beside dual-logic approach (cloud and home logic) introduces a unique way of simulating neural

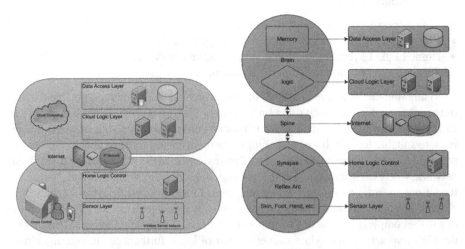

Fig. 1. SHS overall system architecture   **Fig. 2.** Reflex Arc in SHS

pathways that can control action reflex. In animals which are vertebrates, neural signals do not transmit directly into the brain, instead synapse in the spinal cord is used to redirect the signal; for instance, retraction after touching hot objects. This phenomenon is called reflex arc and displays relatively fast reactions in comparison to neural signals which travel all the way to the brain. The proposed architecture is trying to achieve a similar result Fig. 2 illustrates how each component within the four-tier structure corresponds to the human neural reflex arc.

The brain, spine, synapse and tactile organs make up the very basic human neural network, and the layer division is based on this model. Information gathered by sensors is marked as the beginning of the reflex path, some are continuous sensor readings and others are events triggered by humans or the surrounding environment. The data access layer is where all the necessary information is located so that data can be utilized by the logic layer to perform data analysis. Cooperation between data and logic layers forming the equivalent of the human brain that is capable of extracting certain knowledge for decision making. Decisions from this artificial brain are then delivered through the internet infrastructure, which is the same as the spine that links the brain and tactile organs. However unlike human brains (which each brain gives different personalities), there is only one entity in the SHS system and every individual household is attached to it. This one-to-many control model provides general guidance instructions.

The reflex arc in SHS can be described as changes in the sensor readings. The readings can then be processed and corresponding decisions can be made directly on the home domain independent of cloud logic. Different from the synapse (for the purpose of reducing data traffic as well as server load), home logic is designed to upload the sensor readings that cannot be handled by the reflex arc. Due to the nature of IP packet routing, the connection between a home and cloud is not guaranteed; the response time is also highly dependent on the network quality. Without reflex arc, the sensor data will be first sent to the server to be processed, before transmitting back through the same channel, which results in delaying the response time of the control unit. This not only increases the chance of data loss, but also brings a number of uncertainties to the system. For having faster and more compact solution, the home control unit needs to carry out some simple tasks: the definition of a simple task in SHS is given as a condition driven event. Here are the possible scenarios:

- Time: morning, afternoon, evening or exact time, etc.
- Environment: hot, cold, sunrise, sunset and sunlight intensity, etc.
- Human behaviour: work, holiday, shopping, party and etc.

The SHS reflex arc mainly tries to reduce the traffic and speeding up the response time by making decision based on already known daily routines (repeated patterns), and the above simple tasks can be handled locally by the reflect arc. Information that requires further processing and are more complex will be then transmitted to the server (cloud platform), those information will be analyzed by the server and translated into one of the condition driven categories (predefined to the reflex arc system). Consequently the relevant commands will be produced by the reflex arc and send to the end device. In addition to the ability to schedule data uploading, home logic can also choose to not obey the conditions set by the server; this is due to the variability of an individual household. In general, for the purpose of speeding up the response time and minimizing load on the cloud server, SHS reflex arc and the cloud platform divide system decision making into two separate stages. The first stage contains straightforward tasks that are already known to the local home domain and can be executed without assistance from the server. The second stage involves creating solutions for unrecognized tasks based on feedback from the cloud logic.

3   SHS Signal Beam Approach

Every wireless device has the capability to retrieve its transmission signal strength which is known as Receive Signal Strength Indicator (RSSI). One way of utilizing this is based on attenuation with distance [7, 8]. XBee sensor has been used as the primary module within the wireless network. An experiment is carried out for the purpose of determining whether signal strength produced by the XBee module can reflect distance (between the receiver and sender) inside a building. Two XBee sensors have been used in this experiment which one of them is stationary (sender), and the other one is moved continually away from the sender by a distance of one meter until 29 m. At each meter, 100 samples are collected and accordingly average RSSI value is produced. The result shows (Fig. 3) non-linear relationship between the distance and the signal strength. Which means the RSSI values can be similar in different distance. In order to overcome the drawbacks of signal variation and improve indoor localization, a new approach called SHS signal beam is proposed which is inspired by laser beam technology and can be deployed under the smart home scenario. Wireless signal can be treated like light travelling in every direction in a straight line, where there is an obstruction there will be a decrease in the signal strength. This approach still makes use of RSSI measurements to detect human activity within a home environment however its main focus is on making two improvements: the need for a calibration process and how to deal with the impact created by the surrounding environment.

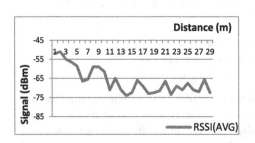

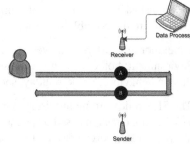

Fig. 3. XBee signal strength variation   **Fig. 4.** Experimental setup

The proposed system needs to undertake some form of a calibration process before deploying. This is due to the strength of the wireless signals which are very sensitive to the changes in the surrounding environment. However, SHS signal beam takes advantage of the interference caused by a wireless signal, and its detection mechanism is based on the amount of variation in RSSI measurements other than the exact value. This method is established based on one hypothesis: the interference effect caused by stationary objects, is constant; which also means that RSSI values should remain stable regardless of presence of any obstacle. The fundamental idea behind this approach is very simple: RSSI values will vary as long as there is human activity within the detection range, this change could either increase or decrease the signal strength. The more movements taking place within a room, the greater the amplitude changes in the signal strength. This can indicate that the RSSI amplitude variation can be used to determine whether a room is occupied or not. To verify this idea the experiment is conducted as illustrated in Fig. 4.

There are two XBee sensors placed line-in-sight with each other, point "A" and point "B" are selected in front of the sensors, where the transfer of wireless signal can get completely blocked by a human body. The purpose of this experiment is to find out how the RSSI value changes, "Before, during and after" a person walks by the points of interest and to examine the feasibility of utilizing those behaviors to detect human activities within a room. The experiment result is illustrated in Fig. 5. A stable output is produced while there is no movement in the room. However signal strength begins to fluctuate slightly when a person approaches the sensors. Points "A" and "B" on the diagram indicate the time when the person is directly infront of the sensor.

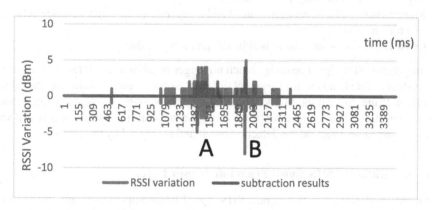

Fig. 5. Processed RSSI variation

The clear straight line at the end of the graph shows that RSSI can recover to its original value right after interference caused by human activity. This initial reading (the straight line) is the reference point in the SHS signal beam approach which defines the condition when human interference does not exist within the range of detection. Although RSSI will be affected by the arrangement of stationary objects inside a room, its reading will stay at a constant level and no calibration process is needed to record this initial condition. This is due to the fact that, the SHS signal beam approach pays more attention to variation trends than actual values.

In order to find out in detail how RSSI values correspond to human activity, a simple calculation regarding the RSSI amplitude variation has been implemented during the experiment: the latest RSSI reading is always subtracted by the previous value; subtraction results are plotted to a 2-dimensional coordinate system as the red line in Fig. 5. The following list briefly describe the finding of this experiment:

- Amplitude 0 (dBm): no movement and the system should keep retrieving RSSI value.
- Amplitude 1: if such fluctuations occur continuously, this will be an indication to the system that there are human influences occurring close by.
- Amplitude 2 & 3: human activity that is very close to the signal source.
- Amplitude 4 and above: There is an object just blocking the wireless transmission path between two devices and it is about to pass through the centre line.

Based on above findings, SHS signal beam approach can be divided into two major phases: idle and detecting. An idle phase is where the amplitude value is always lower than 1; the system should switch to detection mode when a continuous reading of greater than 1 is detected; amplitude reading has to be reduced to 0 before the system can switch back to idle again. Apart from the initial condition, several additional conditions are introduced by this approach; a variable is assigned to each condition, which determines when certain events should be triggered:

- Idle condition: 0 is the preference value.
- Start condition: 1 is the preference value; however larger value can be set to exclude any false alarms caused by noise signal.
- Movement Condition: between 1 and 4 and it depends on the actual distance between two transmission devices.
- Cross Condition: 4 and above will be the preference value.

In general, SHS signal beam approach no longer requires a calibration process, and the system is able to adapt to a variety of different environments by adjusting the initial condition value. More importantly, the surrounding environment has much less impact on the detection result comparing to the other methods, as a result of the actual RSSI reading having no effect on the detection result and is ignored by the system.

3.1 Integration of SHS Signal Beam into Sensor Layer

As mentioned earlier in this paper, SHS signal beam approach is not intended to create an independent indoor localization application; which means the SHS functionality is still reliant on existing wireless control modules. The purpose of this approach is to maximize the utilization of existing resources, so that equivalent functionality can be accomplished without adding additional hardware. This not only reduces the cost on hardware equipment, but also minimizes structural changes while deploying the solution. The main priority of this approach is to provide additional sensor information input, so that SHS can take advantage of user indoor location and improve people's quality of life.

The sensor layer in SHS mainly consists of a room box unit and smart sockets (which has wireless module embedded in), because deployment of this layer is based on the actual structure of a room, there is no particular arrangement on how each individual device should be placed, (a mesh network topologic is applied based on the standard

ZigBee protocol, a single data packet may travel through several hops before reaching its destination, depends on the transmit distance). The smart socket is designed to adapt to SHS and it is equipped with the XBee module in order to provide extra functionalities (such as the ability to control appliances remotely) that are not available in a conventional socket. Apart from features such as remote accessibility, the smart socket can also relay the real-time environmental monitoring data to the home logic layer. A reliable data communication channel is required to implement these features and the room box unit is the link between the smart socket and the home logic layer. A gateway device is used to forward sensor data to the collection unit and distributes instructions to each socket unit.

Figure 6 illustrates two deployment scenarios: an ideal configuration and a possible deployment. The ideal deployment has the room box unit placed in the center of the room and each socket is located in the center of surrounding walls; the first reason for this kind of configuration to be called ideal is that the four wireless data transmissions are perpendicular to each other, which naturally divides the whole room into four equal detection sections. Based on the method mentioned in Sect. 3.1 (Integration of SHS signal beam into sensor layer), there are also four sets of RSSI values available for the system to detect the approximate location: as long as "movement condition" occurs in either socket C or D, the system can assume there is some activity taking place within the range of Sect. 1; furthermore, this section will always be the default starting position due to it being located next to the entrance. Another reason for scenario 1 being ideal is that, the distance between each socket is evenly separated, which guarantees that activities occurred within one section will not have a significant impact on more than one detection device, therefore "cross condition" can be utilized by this approach to predict people's movement.

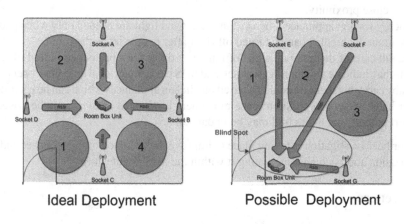

Ideal Deployment   Possible Deployment

Fig. 6. SHS signal beam configurations

The smart socket is designed to be a direct replacement for existing double 13 Amp sockets; this makes the installation process easier as no new wiring is required. In order to fit the ideal configuration, householders would have to carry out some construction work in order to relocate their power sockets which is not ideal. The "Possible deployment" presented on Fig. 6 is an example of how SHS signal beam approach managed to adapt to an existing home environment without the need for relocation of sockets. The room box unit in this layout is placed at the entrance of a room where people can easily gain access to it. Three smart sockets are located at random positions but still manage to produce three detection sections, although this creates a blind spot in the system (a region right in front of the room box unit illustrated with a red circle). Consequently, the system cannot accurately detect a motion because a person would be very close to the receiver node and may completely block the signals. Blind spots would occur when all three transmissions converge together and no individual changes can be detected in the RSSI values. Even though this configuration has some limitations to detect motion within the blind spot area, the RSSI value variation is still able to detect the presence of humans.

4   Conclusion

The new SHS approach has been proposed in this paper which is based on four-tier architecture The fourth tier is added by the author utilizing the "SHS reflect arc" for the purpose of improving the response time of the control unit and better handling of data traffic. By utilizing the wireless sensor network within the SHS architecture, an RSSI based indoor localization technique has been developed; its role is to provide extra sensor input so that more accurate decision can be made by the SHS. A list of its applications is summarized as follows:

- Home automation: the presence of human activities can be the triggers for home appliance; for instance, garage door can be opened automatically when a homeowner is in close proximity.
- Energy saving: appliances such as fan, heater and light are needed only when people are around, and they can be turned off right after people leave the room.
- Health care: if there are no signs of human activity for a prolonged period of time, this may indicate a sudden heart attack and SHS can contact local emergency services.
- Baby care: this scenario could be based on kitchen appliances which can be hazardous for babies or young children; once movement is detected an alarm can be raised to alert parents that their child may be in danger.

Further investigation is taking place to improve the accuracy of the proposed indoor localization approach and integrating it within the SHS architecture.

References

1. Bing, K., Fu, L., Zhuo, Y., Yanlei, L.: Design of an internet of things-based smart home system. In: Proceedings of 2nd International Conference on Intelligent Control and Information Processing, ICICIP 2011, no. PART 2, pp. 921–924 (2011)
2. Helal, S., Hammer, J., Zhang, J., Khushraj, A.: A three-tier architecture for ubiquitous data access. In: Proceedings of ACS/IEEE International Conference on Computer System and Application, pp. 177–180 (2001)

3. Varalakshmi, P., Selvi, S.T., Monica, S., Akilesh, G.: Securing trustworthy three-tier grid architecture with spam filtering. In: Proceedings of 1st International Conference on Emerging Trends in Engineering and Technology, ICETET 2008, pp. 396–399 (2008)
4. Kang, B., Park, S., Member, S., Lee, T., Park, S.: IoT-based monitoring system using tri-level context making model for smart home services. In: IEEE International Conference on Consumer Electronics, pp. 198–199 (2015)
5. Application Highlight and Related Products, "XBee ® & XBee-PRO ® ZB." http://www.digi.com/pdf/ds_xbee_zigbee.pdf. Accessed 27 September 2015
6. Digi, XBee/XBee-PRO ZigBee RF Modules User Guide (2015). http://www.digi.com/
7. Pires, R.P., Gracioli, G., Wanner, L., Frohlich, A.A.M.: Evaluation of an RSSI-based location algorithm for wireless sensor networks. IEEE Lat. Am. Trans. **9**(1), 830–835 (2011)
8. Dian, Z., Kezhong, L., Rui, M.: A precise RFID indoor localization system with sensor network assistance. China Commun. **12**(4), 13–22 (2015)

Threat Navigator: Grouping and Ranking Malicious External Threats to Current and Future Urban Smart Grids

Alexandr Vasenev[1](✉), Lorena Montoya[1], Andrea Ceccarelli[2],
Anhtuan Le[3], and Dan Ionita[1]

[1] University of Twente, 7522 NB Enschede, The Netherlands
{a.vasenev,a.l.montoya,d.ionita}@utwente.nl
[2] University of Florence, Viale Morgagni 65, Firenze, Italy
andrea.ceccarelli@unifi.it
[3] Queen Mary University of London, London E1 4NS, United Kingdom
a.le@qmul.ac.uk

Abstract. Deriving value judgements about threat rankings for large and entangled systems, such as those of urban smart grids, is a challenging task. Suitable approaches should account for multiple threat events posed by different classes of attackers who target system components. Given the complexity of the task, a suitable level of guidance for ranking more relevant and filtering out the less relevant threats is desirable. This requires a method able to distil the list of all possible threat events in a traceable and repeatable manner, given a set of assumptions about the attackers. The *Threat Navigator* proposed in this paper tackles this issue. Attacker profiles are described in terms of *Focus* (linked to *Actor-to-Asset* relations) and *Capabilities* (*Threat-to-Threat* dependencies). The method is demonstrated on a sample urban Smart Grid. The ranked list of threat events obtained is useful for a risk analysis that ultimately aims at finding cost-effective mitigation strategies.

Keywords: Smart grid · Threat assessment · FAIR · NIST · Risk analysis

1 Introduction

Since smart grids are common targets of cyber-attacks [1], there is a clear need to inform smart grid stakeholders about relevant security threats. Moreover, it is of great benefit if a structured ranking of threats is provided.

However, ranking threats for a specific system poses several challenges: (i) first, it is complicated or time-consuming to explore each individual threat in detail in relation to complex smart grids; (ii) second, there is a severe lack of suitable and widely-accepted methods. In particular, although *qualitative* approaches (e.g., built on the NIST 800-30, NISTIR 4628, or NIST 800-53) benefit from advanced categorizations of threats, they comprise only high-level descriptions that do not specify how to achieve the threat ranking goal. A *quantitative* analysis based on the FAIR [2] taxonomy, has the potential to address the challenge, but using it alone may hamper reusing the threat categorization previously mentioned. Therefore, a method for coherently ranking

© ICST Institute for Computer Sciences, Social Informatics and Telecommunications Engineering 2017
J. Hu et al. (Eds.): SmartGIFT 2016, LNICST 175, pp. 184–192, 2017.
DOI: 10.1007/978-3-319-47729-9_19

threats for smart grids using a fusion of these two approaches, e.g., when threats are identified using NIST and ranked using FAIR, would constitute a significant research contribution to smart grids threat assessments. To achieve such an objective, this paper proposes the *Threat Navigator*, which involves a method to filter out less relevant threats, and rank threat events in a rigorous yet flexible manner.

The *Threat Navigator* aims to help stakeholders concentrate on threats with high Loss Event Frequency (*LEF*) in a traceable and repeatable way, thus reducing the number of threat events that need to be further analysed. Input data for the *Threat Navigator* can be extracted from NIST standards with different levels of granularity.

The practical relevance of the *Threat Navigator* relates to that of Intel's TARA methodology [3], specifically, because efficiency and efficacy of cyber-physical risk assessments can be increased if a suitable level of guidance is adopted to distil large amounts of possible attacks into a digest of the most relevant [3].

The method advances the state of the art with regards to other publications in the domain, e.g., [4–6], as it offers the possibility to relate system threats to system assumptions. This allows the rapid investigation of system design alternatives or identification of the most feared attackers (malicious actors) and threats, even when assumptions on attackers or threats change.

In the next sections we outline the adopted threat taxonomy, present the *Threat Navigator* method, and show how it can be applied to urban smart grids.

2 Threat Taxonomy

This section describes the adopted relevant threat factors derived from the mentioned methodologies. These factors are later applied to threat grouping and ranking. We build on the FAIR taxonomy, where *Risk* (the probable frequency and probable magnitude of future loss) is calculated using *Loss Event Frequency* (*LEF*, the frequency, within a given timeframe, that loss is expected to occur) and *Probable Loss Magnitude*. *LEF* is further subdivided into *Threat Event Frequency* and *Vulnerability*. Vulnerability deals with *Control Strength* and *Threat Capability*.

The adopted taxonomy bridges constructs from FAIR and NIST (Fig. 1). We consider NIST's *Targeting* concept to correspond to FAIR's *Contact* concept, while *Intent* corresponds to malicious *Actions*. It enables the use of the standards in a complementary manner, thus benefiting from both of them. This includes the possibility to calculate LEF for a threats' list that stems from the application of NIST 800-30.

In our methodology, we group *Contact* and *Action* (i.e. *Targeting* and *Intent*) into the *Focus* construct. This grouping is reasonable as *Targeting* can be linked to (e.g., reinforced by) *Intent* and both factors are asset-specific. *Focus* depicts how a malicious actor(s) aims at a specific asset (e.g., a type of electric grid element) and indicates the degree of probable physical and cyber contacts of the attacker with the grid assets, together with the attacker intention to launch attacks. *Control Strength* indicates the strength of a counter-measure to limit the success of the attacker.

The adopted threat taxonomy for the needs of this research is therefore composed of *Focus* and *Vulnerability*. The latter includes *Control Strength* as a system characteristic and *Threat Capability* as an attacker characteristic. Together, *Focus* and *Threat*

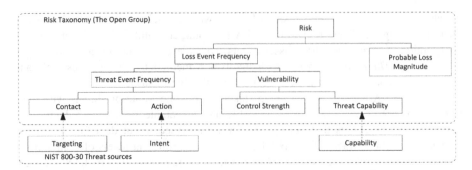

Fig. 1. FAIR risk taxonomy and the adopted NIST-to-FAIR mapping of threat factors.

Capability are two threat factors that depict attacker profiles and constitute a structure to differentiate specific classes of threat actors. These factors are used within the *Threat Navigator* to find and rank relevant threats.

3 *Threat Navigator* as a Method to Rank *LEF* of Threats

The proposed *Threat Navigator* method (Fig. 2) takes pre-processed input data and outputs the *LEF*s of relevant threats.

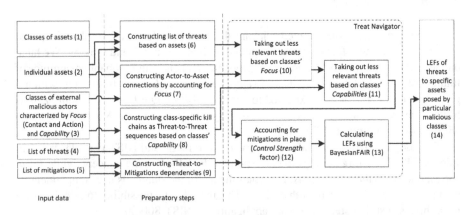

Fig. 2. Proposed *Threat Navigator* method.

The input data are the following: (1) "Classes of assets" are grid assets, (2) individual grid components reflect the categories and individual assets, (3) "Classes of external malicious actors" define which actors apply to individual threats, (4) "List of threats" represents the generic list of threats to be considered; it is later used to identify all threats for the given assets, (5) "List of mitigations" includes controls that can be implemented against the threats listed in [4]. These data are pre-processed to construct (6) a list of threats to be refined, (7) *Actor-to-Asset* and (8) *Threat-to-Threat* connections, and (9) *Threats*-to-*Mitigations* links.

Subsequently, the *Threat Navigator* uses the constructed relations to remove threats less relevant to specific classes of attackers based on their *Focuses* (10) and *Capabilities* (11). Next, the method accounts for implemented mitigations (12) and finally calculates *LEF* for threats (13).

4 An Example of Input Data and Pre-processing Blocks

Input data (1)–(5) and pre-processing (7)–(9) blocks are generic and remain relevant for every grid configuration, while (6) is linked to a particular grid configuration.

(1) Relevant classes of components include Connections, Energy Provider, Building, Data Centre, and Others.

(2) The list of grid components includes: 1. Electricity Connection, 2. Data Connection, 3. Micro Grid Connection, 4. Connection Adapter, 5. Connection Adapter with Energy Transformer, 6. Long-Range Connector, 7. Power Plant 8. Photo Voltaic Energy Generator, 9. Wind Farm, 10. Factory, 11. Stadium, 12. Hospital, 13. Offices, 14. Offices District, 15. Smart Home, 16. Generic Special Building, 17. Basic Data Centre, 18. SCADA, 19. Data and Electricity Storage, 20. EV Charging Point, 21. Access Point.

(3) We consider three basic classes of malicious actors. *Commodity actors* (C1) are opportunistic; most often they do not have organizational support, e.g., recreational hackers, vandals, and sensationalists. *Targeted actors* (C2), e.g. virus writers, crackers, can have some organizational support. *Actors posing advanced persistent threats* (C3) are highly motivated and may include terrorists and actors of nation-states, organized crime, or corporate espionage.

(4) and (5) The list of all of 28 adversarial threats and 13 mitigations identified using NIST references can be found in [7] together with the selection reasoning.

(6) Contrary to other input data and pre-processing elements, this block is specific to a grid configuration. Using a system of systems approach, it constructs lists of threats for grid features (one or more grid components) under consideration. This includes threats relevant to (i) individual components, (ii) classes of components, and (iii) groups of components. The specifics of this process are described in [8].

(7) *Actor-to-Asset* connections depict how the classes of attackers are linked to the list of assets (blocks 1 and 2) for the C1-3 threats classes (block 3) based on their *Focus* characteristic. Three sets of such mappings are constructed [8]: (a) C1_set {1, 2, 11–16, 20, 21}, (b) C2_set {1–6, 8–17, 19–21} and C3_set {1–21}. Only Power Plant and SCADA components are linked exclusively to class C3. This is because these assets remain significantly less available to insufficiently organized actors.

(8) *Threat-to-Threat* sequences are patterns of attacks i.e., kill chains [9], which vary across different actors. NIST-based threats can be constructed as a kill chain based on threat categories [8] e.g., Perform Reconnaissance and Gather Information (PRGI), Craft or Create Attack Tools (CCAT), Deliver/Insert/ Install Malicious Capabilities (DIIMC), Exploit and Compromise (EC),

Conduct an Attack (CA, i.e. direct/coordinate attack tools or activities), Achieve Results (AR, i.e., cause adverse impacts, obtain information), and Coordinate a Campaign (CC). This generic kill chain structure can be populated for classes C1–C3 as follows (number in parenthesis indicate the threat number):

C1: PRGI (2)-CCAT(4)-EC(9,11,12)-CA(17,18)-AR(23–25)

C2: PRGI(1–3)-CCAT(4)-DIIMC(6)-EC(8–12,14)-CA(15–18, 20–22)-AR (23–25)+CC (26)

C3: PRGI(1–3)-CCAT(4)-DIIMC(5–7)-EC(8–14)-CA(15–22)-AR(23–25) +CC (26–28)

These sequences represent *Threat-to-Threat* connections. They include interrelations between (i) individual threats, (ii) individual threats and threat categories, and (iii) categories of threats.

(9) *Threat-to-Mitigation* links are described in [7]. Essentially, these are many-to-many relations, where every mitigation is linked to several threats.

5 An Example of Applying the Threat Navigator

In this section the calculation of the *LEF*s of threats relevant to a Factory feature (Factory and Long-Range connector components) within an urban smart grid is illustrated. We assume that the threats (posed by all attackers) identified in block 6 are {5, 13, 22, 23, 30, 36, 37, 2, 3, 7, 8, 12, 14, 19, 26, 28, 32, 33, 34, 35, 1, 6, 15}.

(10) According to the *Focus* property, one can identify that both Factory and the Long-Range connector are linked to attacker classes C2 and C3, hence the sets of threats C2_set and C3_set are the ones to be considered.

(11) Filtering based on the Capability of threat actors is performed, thus filtering Factory threats with respect to attacker classes C2, and C3. Table 1 illustrates threats: (1) Related to a particular attacker class (denoted as "X" in the following table); (2) Not attributed to a specific class (marked as "|"); and (3) Absent in the list of threats of this feature ("−"), although it should be considered in principle, given the Capability of threat actors.

Thus, the list of relevant threats to the feature includes:

- For C2: C2_Factory_feature_threats = {1, 2, 3, 6, 8, 12, 14, 15, 22, 23, 26};
- For C3: C3_Factory_feature_threats = C3_Factory_feature_threats + ΔC2–C3, where ΔC2–C3 corresponds to threats related to C3 but less to C2. The Factory feature list includes threats {5, 7, 13, 19, 28}.

(12) Mitigations for each threat taken from (9) can be grouped according to C2 or C3, as shown in the first two columns of Table 2. The result of analysing which actors can pose threats to specific grid features provides: (1) A list of threats relevant to C2 actors for the grid feature, including threats that apply if a more advanced class should be taken into account; (2) A list of mitigations relevant to these lists of threats. This provides a checklist of mitigations that can be implemented to make the feature more robust to attacks related to a specific class.

Table 1. Threats relevant to the factory feature.

Steps of kill chain	Considered threats	Relevance to class C1	Relevance to class C2	Relevance to class C3
PRGI	1/2/3	|	X	X
CCAT	4	–	–	–
DIIMC	5/7	|	|	X
	6	|	X	X
EC	8	|	X	X
	9/10/11	–	–	–
	12	X	X	X
	13	|	|	X
	14	|	X	X
CA	15/16/22	|	X	X
	17/18	–	–	–
	19	|	|	X
	20/21	|	–	–
AR	23	X	X	X
	24/25	–	–	–
CC	26/27	|	|	–
	28	|	|	X

Table 2. Identifying the degree of implemented controls as a FAIR construct for C2 threats.

Threat number	Relevant mitigations	% of mitigations implemented	Qualitative characterization of controls
1	11, 12, 18	0 %	Very Low
2	11	0 %	Very Low
3	4, 12, 16, 19	25 %	Low
6	4, 17, 19	33 %	Medium
8	1, 4, 12, 15	25 %	Low
12	8, 10, 13, 16	0 %	Very Low
14	4, 5, 19	33 %	Medium
15	2, 12, 18	0 %	Very Low
22	1, 8, 11, 19	0 %	Very Low
23	1, 8, 10, 11, 13, 19	0 %	Very Low
26	4, 9, 10, 12	25 %	Low

In this example it is assumed that mitigation number 4 (Security Assessment and Authorization) was implemented for the list of identified threats of the Factory feature. Essentially, as interrelations between the threats and mitigations are intricate, implementing an individual mitigation may give rise to input vector changes for several threats. The update concerns controls to threats {3, 6, 8, 14, 26} for class C2 (table below). Similarly, we can find the change in controls for threat {7, 19, 28} relevant to the transition from C2 to C3.

Thus, controls for several threats {3, 8, 26, 7, 19, 28} are improved from 0 to 25 %, while for threats {6, 14} the control increased to 33 %. Because of these changes, the *LEF* of each threat changes.

(13) Bayesian FAIR [10] is then used to calculate *LEF* of [*Contact, Action, Threat Capability, Control Strength*] of an individual threat. The vector elements are within the range of qualitative values from 'Very Low' to 'Very High'. This approach is described in [8]. The first three elements of the vector are known from attacker profiles and the last one is derived based on the mitigations implemented. For example, without mitigations, the input vector for C2 threats {1, 2, 3, 6, 8, 12, 14, 15, 22, 23, 26} is [Medium, Medium, Medium, Very Low]. The logic of forming the input vectors for calculating *LEF*s is shown in Table 3.

Table 3. Operationalizing threat parameters as FAIR constructs.

	Contact (FAIR concept)	Action (FAIR concept)	Threat capability	Control strength
C1	Low	Low	Low	% of implemented controls
C2	Medium	Medium	Medium	
C3	High	High	High	

(14) *LEF*s of threats relevant to class C2 calculated using Bayesian FAIR are:
   1. For threats without mitigations {1, 2, 12, 15, 22, 23, 26}, input data corresponds to [Medium, Medium, Medium, Very Low]. The obtained probability of the *LEF* vector [Very Low; Low; Medium; High; Very High] is [0.013; 0.045; 0.503; 0.265; 0.174]. The value of *LEF* is 701.9.
   2. For threats with 25 % increase in mitigations {3, 6, 8, 14, 26}, the input is [Medium, Medium, Medium, Low]. The *LEF* vector is [0.013; 0.045; 0.503; 0.286; 0.153] with *LEF* value 685.1.
   3. For threats with 33 % increase in mitigations {6, 14}, the input is [Medium, Medium, Medium, Medium]. The *LEF* probability vector is [0.013; 0.045; 0.545; 0.265; 0.132] with *LEF* = 651.5.

Analysis of output vectors for C2 threats and their *LEF* values suggests that the *LEF* value decreases non-linearly (e.g., the decrease in *LEF* for threats with only one mitigation implemented is found to be 17). Two mitigations lower a *LEF* by 50. Figure 3 shows *LEF* values of C2- and C3-relevant threat events calculated in the same

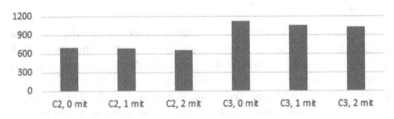

Fig. 3. Ranking groups of threat events.

manner. Potentially, some absolute or relative *LEF* values can be targets for stake-holders who seek to introduce balanced and sufficient mitigations with respect to the threat landscape.

6 Conclusions

This paper proposes the *Threat Navigator* which provides a method for (a) filtering out less relevant threats for a complex system and (b) ranking the rest of the threats. The threat taxonomy adopted within the method bridges the NIST and FAIR methodologies, thus benefiting from both of them, including the calculation of *LEF* for NIST-specified threats using the FAIR taxonomy.

The presented approach is structured but is also highly adjustable, as changes in input data update the consequent filtering and ranking steps in a traceable manner.

Unfortunately, due to the page limit this article cannot fully illustrate the degree of the flexibility embedded into the method, such as: (1) the possibility to account for more attacker classes by applying the briefly outlined profiling approach and (2) specifying the *Control Strength* value increase as a non-linear function of implemented Mitigations. Similarly, other topics were not described in the paper e.g., positioning of the method within the wider assessment process, some specifics of attackers for urban smart grids, and illustrations on how the method can be applied to more sophisticated examples. Further research includes evaluating the extent to which it increases efficiency of cyber-physical risk assessments in comparison to e.g., Intel's TARA. Nevertheless, the *Threat Navigator*'s modular structure suggests that the mentioned modifications and the integration of the method into a risk assessment is plausible.

Ultimately, the aim of this asset-based approach is to focus the risk assessment on what is most relevant. From the point of view of data processing, it avoids the need to manually update vast amounts of data. Thus, smart grid stakeholders can rapidly consider relevant threats and related countermeasures for multiple threats simultaneously. Together with impact analysis, it can therefore help to determine the most cost-efficient mitigation strategies to be implemented.

Acknowledgments. This work has been partially supported by the Joint Program Initiative (JPI) Urban Europe via the IRENE project. We would like to thank Prof. Roel Wieringa for his valuable contribution.

References

1. CS-CERT: ICS-CERT year in review. https://ics-cert.us-cert.gov/sites/default/files/Annual_Reports/Year_in_Review_FY2014_Final.pdf
2. The Open Group: Technical standard. Risk taxonomy. http://pubs.opengroup.org/onlinepubs/9699919899/toc.pdf
3. Intel IT: Prioritizing information security risks with threat agent risk assessment. http://www.intel.com/Assets/en_US/PDF/whitepaper/wp_IT_Security_RiskAssessment.pdf

4. Najgebauer, A., Antkiewicz, R., Chmielewski, M., Kasprzyk, R.: The prediction of terrorist threat on the basis of semantic association acquisition and complex network evolution. J. Telecommun. Inf. Technol. **2008**, 14–20 (2008)
5. Lund, M.S., Solhaug, B., Stølen, K.: Risk analysis of changing and evolving systems using CORAS. In: Aldini, A., Gorrieri, R. (eds.) FOSAD 2011. LNCS, vol. 6858, pp. 231–274. Springer, Heidelberg (2011). doi:10.1007/978-3-642-23082-0_9
6. Morison, K., Wang, L., Kundur, P.: Power system security assessment. IEEE Power Energy Mag. **2**(5), 30–39 (2004)
7. IRENE: D2.1 threats identification and ranking. http://www.ireneproject.eu
8. IRENE: D2.2 societal impact of attacks and attack motivations. http://www.ireneproject.eu
9. Hutchins, E.M., Cloppert, M.J., Amin, R.M.: Intelligence-Driven Computer Network Defense Informed by Analysis of Adversary Campaigns and Intrusion Kill Chains, p. 3. Lockheed Martin Corporation, Bethesda (2010)
10. Le, A., Chen, Y., Chai, M., Vasenev, A., Montoya, L: Assessing loss event frequencies of smart grid cyber threats: encoding flexibility into FAIR using bayesian network approach, smartgifts conference on smart grid inspired future technologies (2016)

Invited Papers Track

A New Dynamic Weight-Based Energy Efficient Algorithm for Sensor Networks

Alsnousi Essa, Ahmed Y. Al-Dubai$^{(\boxtimes)}$, Imed Romdhani,
and Mohamed A. Eshaftri

School of Computing, Edinburgh Napier University, 10 Colinton Road,
Edinburgh EH10 5DT, UK
{a.essa,a.al-dubai,i.romdhani,
m.eshaftri}@napier.ac.uk

Abstract. Since sensor nodes have limited energy resources, prolonging network lifetime and improving scalability are essential elements in energy- efficient Wireless Sensor Networks (WSNs). Most existing approaches consider the residual energy of a single node when electing a cluster head (CH), omitting other factors associated with the node, such as its location within the WSN topology and its nodal degree. Thus, this paper proposes a new Dynamic Weight Clustering based Algorithm (DWCA) for WSNs to reduce the overall energy consumption, balance the energy consumption among all nodes and improve the network scalability. The study has examined the performance of the proposed DWCA algorithm using simulation experiments. We compare the performance of our DWCA against some counterparts. The results demonstrate that our algorithm outperforms its counterparts in terms of energy efficiency and scalability.

Keywords: Wireless sensor network · Routing · Topology · Cluster head · Energy-efficiency

1 Introduction

Wireless Sensor Networks (WSNs) have attracted a lot of attention from the research community due to their capabilities and efficient deployment in many fields and applications. Sensor nodes can be deployed in extremely hostile environments to form networks for different types of applications [1]. A WSN consists of a number of sensor nodes, which can communicate with each other with no infrastructure requirement. However, sensor nodes are usually constrained devices in terms of limited memory, battery and processor capabilities [2, 3]. These limitations represent a critical challenge to implement a typical routing protocol for WSNs. Many contributions have been made to overcome sensor node limitations.

A key technique for reducing energy consumption for WSNs is the node clustering, in which the network is divided into a number of sub-networks called clusters [4, 5]. Then, a suitable node is elected for each cluster based on certain criteria to act as a cluster head (CH) or cluster leader, so that the remaining nodes will join the elected CH to form a local network. Each CH will act as data aggregator and forwarder for its cluster

© ICST Institute for Computer Sciences, Social Informatics and Telecommunications Engineering 2017
J. Hu et al. (Eds.): SmartGIFT 2016, LNICST 175, pp. 195–203, 2017.
DOI: 10.1007/978-3-319-47729-9_20

to the base station. Electing the CH was under the umbrellas of many schools of thoughts and strategies over the past several years [6]. In fact, clustering has been attractive due to a number of advantages; such as, reducing the routing table size stored in each node, conserving communication bandwidth by avoiding the exchange of redundant messages and isolating the routing changes from one cluster to another [5, 7].

Traditional clustering algorithms elect a CH based on the nodes ID or location information, and involve frequent broadcasting of control messages [8]. It is worth indicating that despite many works on selecting the CHs, this issue is still problematic in WSNs, i.e. The existing protocols suffer from some limitations in electing CH as they are generally not considering certain aspects of system functionality and do not deliver an optimal solution with respect to energy efficiency. Motivated by these observations, a novel energy efficient protocol named "A Dynamic Weighted Clustering algorithm" (DWCA) is proposed in this study which considers the problem of electing CH and improves the efficiency of clustering and a network's lifetime. Unlike the previous studies, the DWCA is based on combining key elements, including energy, location, and distance from the base station. Our simulation experiments confirm that DWCA can outperform its counterparts in terms of efficiency and energy consumption.

The paper is organised as follows. Section 2 we review some related work in WSNs. Section 3 accommodates the problem statement we consider. Section 4 presents DWCA algorithm, Sect. 5 presents the simulator settings and simulation results. Section 6 gives conclusions and future work.

2 Related Work

Dividing a WSN into clusters has been widely investigated by the research community; however, electing the optimal CHs remains an open and challenging issue. Several CH election algorithms have been proposed.

Heinzelman et al., proposed LEACH protocol [9] for sensor networks. It is one of the earlier clustering algorithm in WSNs wherein each node can be a CH and it is assigned a certain probability per round. In other words, a set of sensor nodes are chosen as CHs randomly in the network for each round. After a certain time, the CH election rotates among other nodes in the network with the aim of efficiently utilising the network energy. However, the election of CHs in LEACH protocol is based on a probabilistic approach that does not take into account any extra criteria, which can lead to inefficient energy consumption and unbalanced communication. Moreover, when electing CHs on the basis of probability with non-uniform deployment of nodes there is a high possibility of having elected CHs that are placed in one area of the deployment field, leaving some isolated and orphan nodes without a CH to join.

Younis, presented HEED protocol for CH election [10]. HEED protocol aims to prolonging the network lifetime. It differs from LEACH protocol in the way it elects the CHs. In fact, the election of CHs in HEED protocol is based on two main parameters: residual energy of each node and intra-cluster "communication cost". In HEED, the nodes elected as CHs have higher average residual energy compared to cluster member nodes. However, in HEED protocol the energy consumption is not balanced as the number of CHs generated is very high. Furthermore, the process of electing the CHs creates massive overhead due to a number of iterations required to form clusters.

Chengfa Li, proposed EEUC [11]. EEUC intends to address the problem of draining the batteries of the CH, which is closer to the BS due to heavy relay traffic forwarding from other CHs. EEUC proposed a mechanism for partitioning the nodes into clusters of unequal size where the clusters near the BS would be smaller than those located further from the BS. As a result, the CHs closer to the base station could preserve some energy for inter-cluster data forwarding. The selection of a CH in EEUC is primarily based on the residual energy of each node. However, elected CH near base station are not guaranteed to have more residual energy than those far away from the BS.

Eshaftri et al., proposed the Load-balancing Clustering Based protocol (LCP), [12]. LCP, elects the CH node with the highest energy in the first round, and then rotates the election process inside the cluster itself. The LCP protocol is similar to the HEED protocol in the setup phase; however, it differs in the steady phase. Thanks to the new rotation phase, the re-election of the new CH would occur inside the cluster rather than re-clustering the entire network. Once all the nodes inside the cluster have performed the CH task, the last CH sends a message to the BS to re-cluster the network. Despite its advantages, LCP incurs a delay to re-cluster the network where small clusters could report to the BS to re-cluster the network before the bigger clusters complete their rotate cycle. Compared to other protocols, LCP relies also on the residual energy to select CHs.

3 Problem Statement

Many existing studies have proposed different methods for grouping sensor nodes into clusters [4, 9, 10, 12–14]. A number of previous algorithms have concentrated on selecting the CH by choosing a number of random nodes to be as CH in each round. Furthermore, other algorithm as in [15] adopted a probabilistic approach without considering nodes characteristic, which can result in selecting a CH that has less energy, heavily loaded, or has been elected before. Selecting only the nodes with high energy level has been proved as a suboptimal solution in selecting a CH where the deployment of node is non-uniform [16]. If we consider the scenario illustrated by Fig. 1 and we assume the nodes with highest energy are 1, 2 and 3, the node number 1

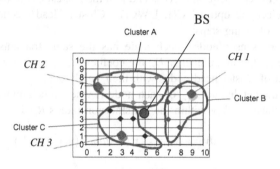

Fig. 1. Potential cluster heads

will become a CH for its set of nodes as it has the highest energy and all the nodes will join as cluster members. However, the node 1 off-centre and even in the edge of the network. Another gap that has not been considered in weighed based algorithms in WSNs is the location of the node and the distance to the base station. If we assume that nodes number 1 or 2 have the highest weight, they will become CH for their set of nodes, where they are in the edge of the network and they have higher cost to communicate with base station.

The algorithm proposed by this paper differs from the aforementioned algorithms in the method of selecting the CH. In fact, we introduced a combined weight for each node to select the CH. The 'Combined weight' is a composite metric that combines energy, neighbourhood, and distance from the base station.

4 Proposed Algorithm

The new Dynamic Weight Clustering Algorithm.

The proposed algorithm attempts to offer the best CH election process in terms of energy efficiency and network lifetime. The new DWCA algorithm is composed of two phases: a setup phase and a steady phase. The election process is similar to most known clustering algorithms for WSNs, as in LEACH and HEED except we have added main critical parameters to elect the CH, this includes the number of neighbours "degree D", the remaining energy "E" and the distance from the BS. Each node calculates its own weight and broadcasts this information to its neighbours. Once the node has received all the neighbouring weights, it will perform a comparison with its own weight. If the node is among the highest weight in the network, it becomes a potential candidate and set the CH candidate flag to TRUE "$CHcandidate = true$".

All CH candidates will broadcast a message, if they meet the condition of highest weight nodes in the network. Only other CH candidates in the network will receive this message to finalize the CH competition process. The message will contain the node ID and the location of the node. The nodes that receive this message will use the competition range function, R_{comp}, to check if there are no other CHs within the competition diameter. The final stage of the competition process is to check whether there are two CH candidate nodes, which are able to hear each other. If so, they will compare their weights and the node with the highest weight will become a final CH, and the other node will give up the competition process and join the nearest CH in the current round.

In order to elect the optimal CH, DWCA's Cluster Head election procedure is composed of the following steps:

Input: A set of sensor nodes, each node has the same transmission radius R_v, transmission rate r_v, the initial energy E_v, the coefficients of weight w1 to w3.

Output: A set of cluster head nodes with their neighbours.

STEP 1: Find the neighbours $N_{(v)}$ of each node v, where a neighbour is a node with its distance with v within the transmission radius R_v. That is:

$$dv = N_{(v)} = \{v' | distance \, (v, v') \leq R_v\},$$

Where d_v is the degree of the node

STEP 2:   Find the distance of $N_{(v)}$ from Base Station $BsD_{(v)}$

$$BsD(v, v') = \sqrt{((x_v - x_{v'})^2 + (y_v - y_{v'})^2)}$$

STEP 3:   The remaining energy E_v is calculated. It indicates how much residual energy a node still hold.

STEP 4:   Calculate the combined weight for every node v

$$W_v = w_1 d_v + w_2 BsD_v + w_3 E_v$$

Where $w_1 = 0.2$, $w_2 = 0.2$, $w_3 = 0.6$

$$w_1 + w_2 + w_3 = 1$$

STEP 5:   Choose the node with a maximum W_V as the CH candidate.

The DWCA algorithm consists of the following phases:

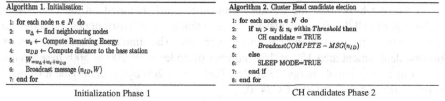

Algorithm 1. Initialisation:
1: for each node $n \in N$ do
2:    $w_\Delta \leftarrow$ find neighbouring nodes
3:    $w_\varepsilon \leftarrow$ Compute Remaining Energy
4:    $w_{DB} \leftarrow$ Compute distance to the base station
5:    $W_{=w_\Delta + w_\varepsilon + w_{DB}}$
6:    Broadcast message (n_{ID}, W)
7: end for

Initialization Phase 1

Algorithm 2. Cluster Head candidate election
1: for each node $n \in N$ do
2:    if $w_i > w_j$ & n_i within *Threshold* then
3:        CH candidate = TRUE
4:        $BroadcastCOMPETE - MSG(n_{ID})$
5:    else
6:        SLEEP MODE=TRUE
7:    end if
8: end for

CH candidates Phase 2

Algorithm 3. Final Cluster Head election
1: On receiving COMPETE-MSG
2: if $(d(n_i, n_j) < n_j.R_{comp}$ OR $d(n_i, n_j) < n_i.R_{comp})$ then
3:    Add s_j to $s_i.S_{CH}$
4: end if
5: while $CHcand = TRUE$ do
6:    if $w_i > w_i$ then
7:        IsCH =TRUE then
8:            Broadcast CLUSTER HEAD-MSG (n_{ID})
9:        else CHcand = FALSE
10:            EXIT
11:    end if
12: end while

Finalization phase 3

1. The initialization phase: each node will listen to *HELLO* messages broadcasted from other nodes to check the number of neighbours. Then the node will broadcast a message stating the *Remaining Energy* and distance to the *BS*.
   Once all the nodes have received the *Hello* messages from their surrounding neighbours, then each node will calculate its *Wight W*.
2. The CH candidates election phase: in this phase after receiving all the W from the nodes n, each node will check its own weight, and if the weight of $n_i > n_j$ and n_i within the *threshold*, then the node will become a CH candidate. The *threshold* is defined as the best 20% of the nodes in the network. If we assume we have 200 nodes in the network and we need to select the best possible nodes as CHs, then

using the algorithm we sort the nodes from highest weight to the lowest, and then select the first 40 nodes (20% of 200 nodes) to compete to become the final CH. The nodes that satisfy the criteria for CH candidates will declare themselves as candidates. All the CH candidates will broadcast *COMPETE-MSG* message with their IDs. Those, which failed to pass the condition, will enter sleep mode until the end of the competition process.

3. Finalization phase: In this phase, the candidate nodes receive the COMPETE-MSG message from other CH candidate. Each node will use the R_{comp} function as in line 2 of Algorithm 3 to check its diameter in comparison with other candidates. If there are no other candidates, the node will become the final CH, otherwise the node with the greatest weigh will become the final CH and the others nodes will stop competing even if they share the same diameter. Once all CH declare themselves the process of forming cluster is similar as in HEED and LEACH protocols.

5   Performance Evaluation

In order to evaluate the performance of DWCA protocol we have carried out several simulation experiments. We set the lifetime of the network as the main performance parameter. We have used the Castalia simulator, which is based on the OMNeT++ platform to run all the experiments. We assume that a set of sensor nodes are deployed randomly throughout a two dimensional square field. The simulation parameters are as follows, deployment area 150 × 150, number of nodes from 200 to 500; base station located in centre, initial energy 25 Joules, Packet size 200 bytes, deployment of node non-uniform and radio model CC2420.

We made the following assumptions for our scenarios:

1- Each node has limited energy (non-rechargeable battery).
2- The nodes are location-unaware.
3- The network topology is static throughout the network lifetime.
4- All the nodes are able to communicate directly with the BS.
5- The BS is located in the centre of the field.
6- All the nodes are homogeneous, where they have same initial energy and transmission range.

The network lifetime is measured using the following three metrics.

First Node Die (FND) is the time elapsed in rounds until the first node has had all of its energy drained. Half Nodes Die (HND) is the time elapsed in **rounds until half** of the nodes have consumed their energy. Last Node Dies (LND), is the time elapsed in rounds until all the nodes have had their total energy drained.

The "round" definition in our paper refers to the time interval in seconds before the network starts a new clustering process. Therefore, there is no difference between the round concepts in DWCA and LCP, EEUC and the original HEED, in terms of time.

Figure 2 illustrates a comparison between DWCA and the three closest clustering algorithms. Figure 2 illustrates the lifetime of 200 nodes. It is evident that the DWCA performs better as the total lifetime has been extended; compared with LCP 12.5%,

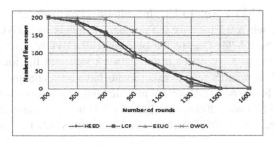

Fig. 2. Number of alive nodes VS number of rounds for HEED, LCP, EEUC and DLCP

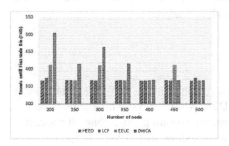

Fig. 3. First node dies in HEED, LCP, EEUC and DWCA.

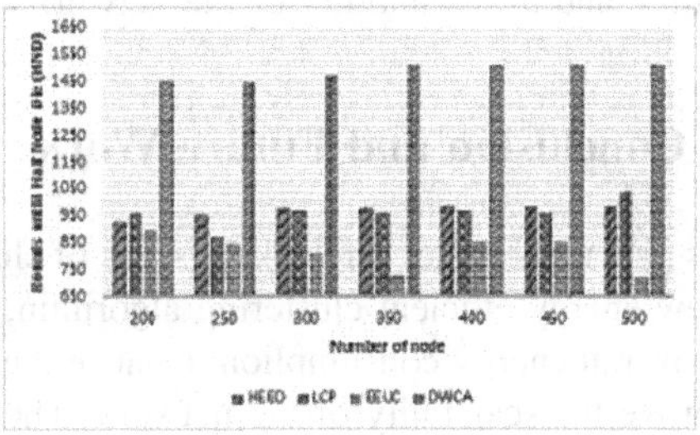

Fig. 4. Half node dies in HEED, LCP, EEUC and DWCA

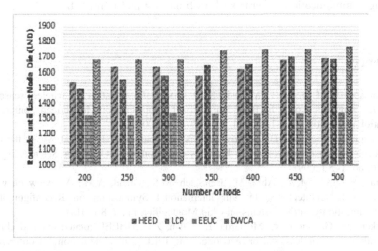

Fig. 5. Last node dies in HEED. LCP, EEUC and DWCA

HEED 9.57%, and 27.31% with EEUC. In addition, DWCA shows stability in term of the nodes that die when we increase the number of rounds and DWCA improves the network lifetime by 9.57%. One of the objectives of DWCA is to prolong network lifetime, the number of alive nodes in the network indicates the lifetime time of the network. Figure 3 shows that several experiments have been conducted using different number of nodes. From the Fig. 3 we can note that our protocol performs better in term of firs node dies FND with 200 nodes. FND in 200 nodes is significantly high compared with other protocols. DWCA shows that it performs better in most the scenarios in term of FND.

Figure 4 indicates that substantial improvement has been gained by the DWCA protocol in term of half node dies HND. In Fig. 4 HEED, LCP and EEUC have lost half of their nodes in just about the half time of DWCA. DWCA outperformers in term of HND in all scenarios. In Fig. 5, it is apparent that the DWCA has improved the total lifetime and the scalability of the network in all scenarios.

6   Conclusion and Future Work

This paper elaborates on the key issues of clustering approaches in WSNs. It introduces a new energy efficient clustering algorithm, DWCA for WSNs, which aims to reduce the overall energy consumption, balance the energy consumption among all nodes and improve the scalability of the network. The results show that DWCA outperforms its counterparts. The method used in DWCA is different from existing works where we used lightweight matrix and critical for CH tasks that allows us to reduce energy consumption and offer better load balancing. This approach can be applied in different applications with respect to system parameters. The future work of DWCA is to investigate the multi-hop manner and examine the proposed DWCA under different operating communication patterns such as broadcast and multicast.

References

1. Arampatzis, T., Lygeros, J., Manesis, S.: A survey of applications of wireless sensors and wireless sensor networks. In: IEEE International Symposium on Mediterrean Conference on Control and Automation, Limassol, pp. 719–724 (2005)
2. Kanavalli, A., Sserubiri, D., Shenoy, P.D., Venugopal, K.R., Patnaik, L.M.: A flat routing protocol for sensor networks. In: Methods and Models in Computer Science, ICM2CS, Delhi, pp. 1–5 (2009)
3. Karray, F., Jmal, M.W., Abid, M., BenSaleh, M.S., Obeid, A.M.: A review on wireless sensor node architectures. In: 9th International Symposium on Reconfigurable and Communication-Centric Systems-on-Chip Montpellier, pp. 1–8 (2014)
4. Aierken, N., Gagliardi, R., Mostarda, L., Ullah, Z.: RUHEED-rotated unequal clustering algorithm for wireless sensor networks. In: IEEE 29th International Conference on Advanced Information Networking and Applications Workshops (WAINA), Gwangiu, pp. 170–174 (2015)

5. Singh, J., Kumar, R., Mishra, A.K.: Clustering algorithms for wireless sensor networks: a review. In: 2nd International Conference on Computing for Sustainable Global Development (INDIACom), New Delhi, pp. 637–642 (2015)
6. Boyinbode, O., Le, H., Mbogho, A., Takizawa, M., Poliah, R.: A survey on clustering algorithms for wireless sensor networks. In: 13th International Conference on Network-Based Information Systems (NBiS), Takayama, pp. 358–364 (2010)
7. Patil, H.D., Sugur, S.M.: Energy efficient clustering algorithm in WSN. In: 2nd International Conference on Signal Process Integration networks, vol. 2, pp. 1–10 (2015)
8. Modi, A., Prajapati, G.: WSN performance issues and various clustering methods. Int. J. Comput. Appl. 107(15), 28–34 (2014)
9. Handy, M.J., Haase, M., Timmermann, D.: Low energy adaptive clustering hierarchy with deterministic cluster-head selection. In: 4th International Workshop on Mobile and Wireless Communications Network, pp. 368–372 (2002)
10. Younis, O., Fahmy, S.: HEED: a hybrid, energy-efficient, distributed clustering approach for ad hoc sensor networks. IEEE Trans. Mob. Comput. 3(4), 366–379 (2004)
11. Li, C., Ye, M., Chen, G., Wu, J.: An energy-efficient unequal clustering mechanism for wireless sensor networks. In: IEEE International Conference on Mobile Ad Hoc and Sensor Systems, Washington, DC, p. 604 (2005). Pages 8
12. Eshaftri, M., Al-Dubai, A.Y., Romdhani, I., Yassien, M.B.: A new energy efficient cluster based protocol for wireless sensor networks. In: Federated Conference on Computer Science and Information Systems, Lodz, pp. 1209–1214 (2015)
13. Jarchlo, E.A.: Life Time sensitive weighted clustering on wireless sensor networks. In: Proceedings 3rd International Conference Sensor Networks, pp. 41–51 (2014)
14. Sabri, A., Al-Shqeerat, K.: Hierarchical cluster-based routing protocols for wireless sensor networks–a survey. Int. J. Comput. Sci. IJCSI 11(1), 93–105 (2014)
15. Kour, H., Sharma, A.K.: Hybrid energy efficient distributed protocol for heterogeneous wireless sensor network. Int. J. Comput. Appl. 4(6), 1–5 (2010)
16. Sharma, N., Nayyar, A.: A comprehensive review of cluster based energy efficient routing protocols for wireless sensor networks. IJAIME 3(1), 441–453 (2014)

LTE Delay Assessment for Real-Time Management of Future Smart Grids

Ljupco Jorguseski[✉], Haibin Zhang, Manolis Chrysalos,
Michal Golinski, and Yohan Toh

Department of Network Technology, TNO,
Anna van Buerenplein 1, 2595 DA Den Haag, The Netherlands
{ljupco.jorguseski,haibin.zhang,manolis.chrysalos,
michal.golinski,yohan.toh}@tno.nl

Abstract. This study investigates the feasibility of using Long Term Evolution (LTE), for the real-time state estimation of the smart grids. This enables monitoring and control of future smart grids. The smart grid state estimation requires measurement reports from different nodes in the smart grid and therefore the uplink LTE radio delay performance is selected as key performance indicator. The analysis is conducted for two types of measurement nodes, namely smart meters (SMs) and wide area monitoring and supervision (WAMS) nodes, installed in the (future) smart grids. The SM and WAMS measurements are fundamental input for the real-time state estimation of the smart grid. The LTE delay evaluation approach is via 'snap-shot' system level simulations of an LTE system where the physical resource allocation, modulation and coding scheme selection and retransmissions are modelled. The impact on the LTE delay is analyzed for different granularities of LTE resource allocation, for both urban and suburban environments. The results show that the impact of LTE resource allocation granularity on delay performance is more visible at lower number of nodes per cell. Different environments (with different inter-site distances) have limited impact to the delay performance. In general, it is challenging to reach a target maximum delay of 1 s in realistic LTE deployments (This work is partly funded by the FP7 SUNSEED project, with EC grant agreement no: 619437.).

1 Introduction

One intrinsic property of future smart grids is the distributed energy resources (DER), such as wind turbines and photovoltaics. Due to the expected high number of DER, the power flow in future smart grids migrates from traditional one way flow from major power plants to consumers, towards power flow in both directions. Additionally, it is expected that consumers will utilize electricity demand-response approaches in order to lower their monthly energy cost. Consequently, the voltage violations and to the lesser extend congestions in the electricity distribution network could occur. A traditional approach to deal with these operational issues is to perform the necessary electricity grid reinforcements, and limit with sufficient margin the power flow injected from the DER into the electricity grid, in order to safeguard the necessary voltage levels. However, this increases operational costs and does not fully leverage the potential of

© ICST Institute for Computer Sciences, Social Informatics and Telecommunications Engineering 2017
J. Hu et al. (Eds.): SmartGIFT 2016, LNICST 175, pp. 204–213, 2017.
DOI: 10.1007/978-3-319-47729-9_21

the DER. A more future proof approach, which avoids or delays grid reinforcements and fully utilizes the DER potential, is deployment of advanced distribution network management system platform (ADNMSP) [1]. This platform should enable real-time monitoring of voltage profiles and power flows via so-called wide area monitoring and supervision (WAMS) nodes which perform measurements and generate input for grid state estimations (also including demand and supply forecast), initiate congestion management actions, etc. The whole concept of ADNMSP is presented in detail in [2].

The WAMS nodes should be installed in carefully selected locations that enable optimized state estimation of the electricity distribution grid with minimum number of installed WAMS nodes. Further, the ANDSMP will also utilize the measurements from smart meters (SMs). The WAMS or SM measurements include e.g. voltage, current, power, harmonics, etc. and reliability parameters (e.g. number and duration of outages).

In this paper we address the communication network supporting the measurement reporting for WAMS and SM nodes via uplink transmission in LTE cellular networks.

The LTE system capability, specifically delay/latency performance is investigated for smart grids applications [3–5] as well as in general (e.g. for supporting voice over IP services) [6, 7]. The approaches in these studies involve measurements or experiments complemented by theoretical analysis. The impact on the delay performance from the number of active users in the cell is however not addressed, which is one of the main topic of this study. Further, the study also addresses the impact on the uplink transmission delay from the granularity of the allocated transmission resources for each measurement node as well as the environment type. It is noted that, there have been also studies on the random access channel (RACH) performance in LTE when supporting smart grid applications [8]. This is left out of the scope of this study as it is assumed that WAMS nodes (or even SMs if reporting with low period e.g. below or equal to 1 s) would not perform the RACH procedure before sending individual measurements in uplink but rather have ongoing active sessions with intermediate discontinuous transmission (DTX) cycles.

The rest of the paper is organized as follows. In Sect. 2 the communication requirements and communication modelling approach used in this study are presented. Section 3 presents the analysis approach for the LTE delay performance followed by the numerical simulation results in Sect. 4. The paper is finalized with the conclusions and recommendations in Sect. 5.

2 Smart Grid Communications Requirements and Communication Modelling Assumptions

The WAMS nodes (and SMs) are the basic facilitating nodes for the ADNMSP approach. In this section we aim at defining the communication requirements in terms of amount of generated measurement data and the end-to-end delay requirements for the transmitted measurement reports.

The installation of WAMS nodes in the medium and low voltage parts of the electricity distribution grid is beyond current state of the art. Each WAMS node has to perform and report similar measurements as the so-called phasor measurement unit (PMU) in the high voltage (transmission) electricity grid [9, 10]. Consequently, the size of the reported measurement from these WAMS nodes can be derived from the PMU measurement reporting with floating point representation as follows [10]:

$$22 + \left(n_{ph} \times 8\right) + \left(n_{an} \times 4\right) + \left(n_{dig} \times 2\right) \; [\text{Bytes}]$$

Here, we have a header overhead of 22 bytes, n_{ph} is the number of (e.g. voltage, current, etc.) phasors, n_{an} and n_{dig} represent the number of analog or digital measurements, respectively. Assuming $n_{ph} = 6$ for e.g. voltage and current phasor in a 3-phase system and assuming for simplicity single additional digital measurement, i.e. $n_{an} = 0$ and $n_{dig} = 1$, we arrive at 72 Bytes for each WAMS measurement sample. For the packet size of the SM nodes we assume 56 Bytes as reported in [20].

Next to the measurement report size it is important to know the delay requirement for the reported measurements from the WAMS and SM nodes. The literature gives the following indications:

(1) SM related latency requirements (Advanced Metering Infrastructure): 1 s [12].
(2) WAMS related latency requirements: less than 1 s, typically 100–200 ms [11].
(3) The smart grid industry indicates in [13], among other requirements, that in the access part (low and medium voltage) of the electricity grid the communication between the end-nodes and the secondary substations with distances typically less than 10 km requires a guaranteed end-to-end delay (i.e. strict upper bound) of less than 1 s.

In this study the focus is on the radio part of the LTE uplink transmission delay d_{radio}, as illustrated in Fig. 1. The additional delay in the core network part d_{core}, i.e. from the evolved Node B (eNB) to the point where the transmitted packet enters the Internet domain outside the mobile operator network, is not taken into account. This is because d_{core} is typically much smaller than d_{radio} and it can be considered relatively constant for varying number of terminals communicating with the eNB (e.g. 10 ms as mentioned in [14] or 20 ms in [15]).

In LTE the shortest radio transmission duration is 1 ms, known also as transmission time interval (TTI). In the time-frequency domain, the blocks of 12 contiguous frequency subcarriers (with 180 kHz width) and one time slot of 0.5 ms are defined as Physical Resource Blocks (PRBs). The PRB allocation is done by the scheduler located in the eNB as illustrated in Fig. 1.

In our delay analysis in Sect. 3, in every TTI a number of nodes willing to transmit are scheduled with a fixed number of PRBs. This follows the scheme of fair fixed assignment (FFA) as described in [16]. Depending on the radio conditions (e.g. Signal-to-Interference-plus-Noise Ratio - SINR level) an appropriate modulation and coding scheme is selected for the transmission in the assigned PRBs. This, in turn determines the amount of data (in bits) that can be transferred, and therefore WAMS and SM terminals experiencing different SINR conditions might use one or more TTIs to transfer their measurement reports. Additionally, as the number of active WAMS and SM terminals increases the total number of available PRBs within one TTI is becoming insufficient to serve all the active terminals, resulting in waiting time of one or more TTIs for the WAMS and SM terminals in order to be scheduled for

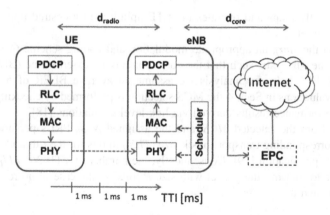

Fig. 1. Illustration of the uplink transmission in LTE and the two delay components d_{radio} and d_{core}. The user equipment (UE) is the 3rd Generation Partnership Project (3GPP) term for end terminals and evolved Node B (eNB) is the term for base stations.

transmission. This consequently increases the delay for the delivery of the measurement report packets.

3 Analysis of LTE Delay and Capacity

In this section, we first exploit the impact of number of nodes on the end-to-end delay performance of WAMS and SMs, by taking into account the communication requirements and assumptions in Sect. 2.

In a certain cell of the LTE network, a number N_{UE} of (WAMS and SM) nodes are typically randomly placed within the coverage area of the cell.[1] For an arbitrary i-th node, the achievable wideband signal-to-interference-plus-noise ratio $sinr_i$ is calculated as follows:

$$sinr_i = \frac{p_{UE}L_iG_{div}}{12N_{PRB}(n_{th\_i} + i_{ul\_i})} \qquad (1)$$

Here, p_{UE} is the uplink transmission power of the node (e.g. 23 dBm). L_i is the propagation loss (incl. antenna gain, path-loss, shadowing, but no multipath fading) for the particular location of the i-th node. G_{div} is the environment-specific macro-diversity gain [17]. N_{PRB} denotes the (fixed) number of PRBs assigned per node, and correspondingly $p_{UE}/12N_{PRB}$ is the transmit power level per allocated frequency subcarrier. n_{th_i} and i_{ul_i} are the thermal noise (including noise figure at the receiver) and inter-cell interference experienced by the i-th node on the allocated PRBs, respectively. For the sake of simplicity, we assume a given uplink inter-cell interference level of $i_{ul\_i} = 3$ dB for all the allocated PRBs. This is motivated by the fact that in the busy hours an

---

[1] UE stands for user equipment.

operator typically plans and operates the LTE uplink with a desired uplink inter-cell interference target.

Based on the $sinr_i$, an appropriate modulation and coding scheme (MCS_i) can be selected for the i-th node, such that the corresponding block-error rate (BLER) is in a range of [0, 10 %]. In our analysis we assume an average BLER of 5 %. For the numerical evaluations in Sect. 4 the MCS selection is performed by checking the SINR vs BLER performance results derived via link-level simulations [18].

Further, from the selected MCS_i and the assigned N_{PRB} PRBs per node, we can derive the corresponding transport block size (TBS_i) in bits for the i-th node, according to the 3GPP specification [19]. Consequently, the number of TTIs Nr_TTI_i needed by the i-th node to transmit the packet with size P [bits] in the case of no retransmission can be calculated as:

$$Nr\_TTI_i = \frac{P}{TBS_i} \tag{2}$$

In order to assess the delay each node experiences, a scheduling simulation loop is executed advancing with a TTI step (i.e. 1 ms step) where at each scheduling turn the following steps are performed:

- Step-1: Determine the number of PRBs which are free for initial transmission allocation, denoted as $N_{PRB,free}$. Ideally, there will be the total number of PRBs of the cell free, e.g. 50 PRBs for a 10 MHz LTE carrier. However, an operator can also decide to reserve only a fraction of the whole LTE carrier for supporting smart grid nodes. From the total free PRBs available for allocation the resources claimed by the following WAMS and SM users have to be extracted:
  - Nodes that were transmitting in previous TTI (if applicable) but need additional TTI to finish the transmission.
  - Nodes that need to re-transmit a packet (if applicable) that was originally transmitted at least 8 TTIs earlier [21][2], see also step-3a below.
- Step-2: Select randomly from at most $N_{PRB,free}/N_{PRB}$ nodes for initial transmission in the given TTI i.e. users that are not continuing their transmission from previous TTI or scheduled for retransmission. Note again that N_{PRB} is the fixed number of PRBs that can be allocated for a single node. For each selected node, the following is performed:
  - Reduce Nr_TTI_i by one as this node is scheduled for initial transmission. If Nr_TTI_i is larger than zero than this node is scheduled for transmission also in the next TTI.
  - If the packet is erroneously received, assuming this will happen in average for 5 % of the transmissions, schedule this node for retransmission at earliest 8 TTIs further ahead [21]. Otherwise set the delay to the current TTI.

---

[2] In LTE the minimum time needed for a transmitter to realize its previous transmission is erroneously received and needs to be re-transmitted is 8 TTIs.

- Step-3a: Select randomly from nodes which are scheduled for retransmission. For each of these nodes, the following is performed (no reduction of Nr_TTI_i as this is a retransmission):
  - If the packet is erroneously received, as this will happen in average for 5 % of the transmissions, schedule this node for retransmission at earliest 8 TTIs further ahead and set the UE delay to the current TTI plus 8 ms. Otherwise set the UE delay to the current TTI.
- Step-3b: Allocation of N_{PRB} resources to nodes that are continuing its transmission from previous (if applicable) TTI. For each selected node, the following is performed:
  - Reduce Nr_TTI_i by one as this node is scheduled for transmission. If Nr_TTI_i is larger than zero than this node is scheduled for transmission also in the next TTI.
  - If the packet is erroneously received, assuming this will happen in average for 5 % of the transmissions, schedule this node for retransmission at 8 TTIs further ahead. Otherwise set the delay to the current TTI.

The scheduling loop is stopped when all nodes have sent their packets including retransmissions. This is because the scope of the analysis is to see what is the upper bound of the delay (i.e. the guaranteed delay) for different number of active WAMS and SM nodes. This can be used to configure the measurement reporting period of the WAMS and SM nodes for 'real-time' smart grid state estimation. For example, if we configure the measurement reporting period to be 1 s, 10 s, etc., then we have to guarantee that all reporting WAMS and SM nodes transfer their measurement packets within the measurement period. Otherwise, the following reporting period will start and the already pending WAMS and SM measurements for transmission become obsolete for the 'real-time' smart grid state estimation.

In this study, within each TTI of the scheduling loop packets from the WAMS nodes are prioritized over the SMs This is to reflect the more importance of WAMS nodes in Smart Grids, and the (to some extent) stricter delay requirement for WAMS nodes [11] relative to the SM nodes [12].

The cell capacity is then defined as the number of WAMS and SM nodes within the LTE cell that can be served such that their respective delay requirements in terms of guaranteed delay are satisfied.

4 Numerical Results and Analysis

In this section, we first present the simulation parameters settings used in our study. Urban and sub-urban environments are considered in the analysis with the most important parameters as specified in Table 1. Numerical results are then presented, following the approach in Sect. 3 above, that quantify the achievable delay and capacity of an LTE network for the smart grid applications.

Simulation Parameters. We assume that each WAMS or SM node is assigned $N_{PRB} = 0.5$, 1 or 2 PRBs in order to investigate the impact of the resource allocation granularity on the delay performance. 3GPP is considering to allow partial PRB allocation even down to single frequency subcarrier for the so-called Narrow Band

Table 1. Simulation parameters.

Parameters	Environment	
	Urban	Suburban
System bandwidth(MHz)	10	10
Inter-Site Distance (m)	500	1000
Spectrum band (MHz)	1800	800
Propagation model	COST231-Hata	Hata-Okumura
Penetration loss (dB)	17	16
Shadowing standard deviation (dB)	10	8
Macro-diversity gain (dB) [17]	5	4
Max. terminal transmit power (dBm)[a]	23	23
Thermal noise power density (dBm/Hz)	−174	−174
Noise figure at eNodeB (dB)	2	2
Antenna gain at devices (dB)	0	0
Antenna gain at eNodeB (dB)	15	15
Receive diversity gain (dB)[b]	3	3
Cable and connector losses at eNodeB (dB)	1	1

[a]Subject to additional losses arising from antenna cables and body loss, which though for a typical communication node in a smart grid deployment should be expected to be absent or negligible.
[b]The theoretical maximum for a configuration with two antennas at the eNodeB.

Internet of Things (NB-IoT) enhancements in LTE. This increases the uplink coverage by increasing the transmit power spectral density as well as reduce the overhead (i.e. increase efficiency) if small amount of data have to be transmitted.

We assume packet size of 56 bytes for SM and 3600 bytes (i.e. 50 samples per measurement report) for WAMS nodes, as discussed in Sect. 2. In the analysis we investigate a 10 MHz LTE system where from the total amount of 50 PRBs only 6, 20 or 50 PRBs are available for allocation to SM and WAMS nodes. This is motivated by the fact that an LTE telecom operator may only utilize a fraction of the total LTE carrier for support of the smart grid nodes.

Finally, at the start of the simulation run as explained in Sect. 2 it is assumed that all active WAMS or SM nodes would like to trigger uplink measurement transmissions uniformly distributed within 1 s interval in order to model the situation in practice when all uplink transmissions from the smart grid nodes does not occur synchronously at the same time instant. The same approach is applicable for other possible time intervals.

Numerical Results and Analysis. The numerical results presented in this section are generated by 'snap-shot' simulations with static nodes as no node mobility is envisaged for the SM and WAMS nodes in smart grid applications. An LTE cell is populated with randomly placed SM and WAMS nodes e.g. from 50 to 2000 nodes per LTE cell. For each 'snap-shot' the scheduling procedure from Sect. 3 is executed and the delay in number of TTIs is determined for sending the packets generated by the individual SM

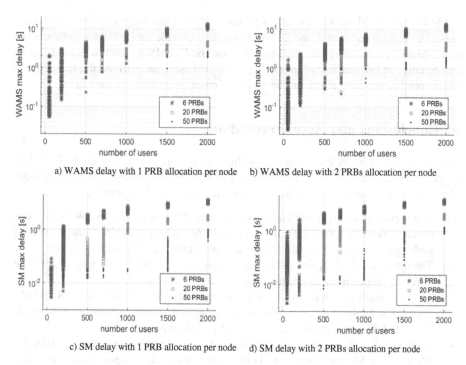

a) WAMS delay with 1 PRB allocation per node   b) WAMS delay with 2 PRBs allocation per node

c) SM delay with 1 PRB allocation per node   d) SM delay with 2 PRBs allocation per node

Fig. 2. Achievable maximum delay of WAMS nodes (top) and SM nodes (bottom) for urban deployment scenario. The WAMS vs SM ratio is 1/3.

and WAMS nodes. The 'snap-shot' realizations are repeated many times (e.g. 1000) and in each realization the individual packet delays are collected forming a statistical set that is used to derive the maximum delay per snapshot as illustrated in the scatter plots in this section. It is important to stress here that all the results in this section should be interpreted for the case of simultaneously active SM and WAMS terminals.

First, we investigate the achievable guaranteed delay for the scenario where each SM or WAMS node is allocated, in a fixed manner, a single PRB per TTI. This analysis is presented in Fig. 2a and c for the case of the urban environment and WAMS to SM ratio of 1/3. As it can be observed from Fig. 2a the maximum delay of WAMS nodes is higher than 1 s e.g. in the range 0.05 to 2 s for low (e.g. 50) number of active nodes. The higher availability of LTE resources for the smart grid traffic (e.g. from 6 to 50 PRBs) does not influence the maximum delay for lower number of nodes due to the fixed 1 PRB allocation per node. The SM maximum delay at low number of users is smaller (Fig. 2c) as these nodes need to transmit much smaller data packets (e.g. factor 64 smaller). Note that due to uniform spread of uplink transmissions instants within 1 s from the WAMS and SM nodes, the SM nodes can still be scheduled for transmissions although WAMS nodes are prioritized due to availability of PRBs to be utilized. As the number of nodes increases the maximum delay of WAMS and SM nodes becomes comparable.

In Fig. 2b and d the same result is presented with fixed assignment of 2 PRBs per node. Comparing between 1 and 2 PRB allocation per node the 2 PRBs assignment per node (higher granularity) has positive effect on the WAMS maximum delay (Fig. 2a and b) only for low number of users (e.g. up to 500 users per cell). At the same time it has somewhat negative effect on the SM maximum delay (Fig. 2c and d).

5 Conclusions and Recommendations

This study investigates the maximum uplink delay performance of an LTE system supporting packet transmission from SM and WAMS nodes in Smart Grids application. The analysis presented in this study leads to the following conclusions:

(a) The impact of different fixed allocated number of PRBs for SM and WAMS nodes on the maximum uplink LTE delay performance is more visible at low number of nodes per cell, while not significant for more than e.g. 500 nodes per cell.

(b) The maximum LTE uplink delay performance is not significantly impacted by the different propagation environment and inter-site distances.

(c) The maximum uplink delay of 1 s for measurement reports from WAMS and SM nodes is difficult to achieve without reserving significant capacity (e.g. a whole 10 MHz LTE carrier) for practical smart grid deployments.

(d) A practical achievable performance would be a maximum delay of 10 s for WAMS and SM measurement. This, hower, should be investigated from smart grid operational point of view if this update interval provides for sufficient 'real-time' observability of the future smart grids.

As a further study the delay analysis will be applied to the extensions of LTE, i.e. LTE-M and NB-IoT, targeted at machine-to-machine (M2 M) applications, to evaluate their relative delay performance with respect to the achievable LTE delays and their suitability for real-time smart grid management. Additionally, other deployment scenarios may be evaluated where SM and WAMS measurements are collected via other RF transmission (e.g. local WiFi or RF-mesh) towards a gateway (e.g. an aggregator node) that is further communicating via a nearby LTE base station. 3GPP is now studying technical solutions to reduce air interface latency, by e.g. reducing the TTI period from 1 ms to 0.5 ms or even one OFDM symbol duration (0.07 ms). This will impact the achievable delay performance of LTE fundamentally.

Acknowledgment. We thank all the colleagues from the FP7 SUNSEED project consortium for the numerous discussions that were useful input for the study presented in this paper.

References

1. http://sunseed-fp7.eu/
2. FP7 SUNSEED, Deliverable D2.1.1: Preliminary requirements and architectures for DSO-telecom converged communication networks in dense DEG smart energy grid networks, July 2014

3. Maskey, N., Horsmanheimo, S., Tuomimaki, L.: Analysis of latency for cellular networks for smart grid in suburban area. In: 5th IEEE PES Innovative Smart Grid Technologies Europe (ISGT Europe), Istanbul, Turkey, 12–15 October 2014

4. Horsmanheimo, S., Maskey, N., Tuomimaki, L.: Feasibility study of utilizing mobile communications for smart grid applications in urban area. In: IEEE International Conference on Smart Grid Communications (SmartGridComm), Venice, Italy, 3–6 November 2014

5. Louvros, S., Paraskevas, M., Triantafyllou, V., Baltagiannis, A.: LTE uplink delay constraints for smart grid applications. In: 19th IEEE International Workshop on Computer Aided Modeling and Design of Communication Links and Networks (CAMAD), Athens, Greece, 1–3 December 2014

6. Nagai, Y., Zhang, L., Okamawari, T., Fujii, T.: Delay performance analysis of LTE in various traffic patterns and radio propagation environments. In: 77th IEEE Vehicular Technology Conference (VTC Spring), Dresden, Germany, 2–5 June 2013

7. Zhang, L., Okamawari, T., Fujii, T.: Performance Evaluation of End-to-End Communication Quality of LTE. In: 75th IEEE Vehicular Technology Conference (VTC Spring), Yokohama, Japan, 6–9 May 2012

8. Karupongsiri, C., Munasinghe, K.S., Jamalipour, A.: Random access issues for smart grid communications in LTE networks. In: 8th International Conference on Signal Processing and Communication Systems (ICSPCS), Gold Coast, Australia, 15–17 December 2014

9. Adamiak, M., Premerlani, W., Kasztenny, B.: Synchrophasors: Definition, Measurements, and Application. https://www.gedigitalenergy.com/

10. Lixia, M.: IEEE 1588 Synchronization in Distributed Measurement Systems for Electric Power Networks, Ph.D. Thesis, University of Cagliari, March 2012

11. Xu, Y., Fischione, C.: Real-time scheduling in LTE for smart grids. In: 2012 5th International Symposium on Communications Control and Signal Processing (ISCCSP), May 2012

12. IEC 61850-5: Communication networks and systems for power utility automation–Part 5: Communication requirements for functions and device models, January 2013

13. 5G Infrastructure Association: "5G and Energy", White Paper, Version 1.0, 30 September 2015

14. Mavenir: Latency Considerations in LTE. White paper, September 2014

15. Nokia Siemens Networks: LTE-capable transport: A quality user experience demands an end-to-end approach, Whitepaper (2011)

16. Dimitrova, D.C., Berg, J.L., Heijenk, G., Litjens, R.: LTE uplink scheduling - flow level analysis. In: Sacchi, C., Bellalta, B., Vinel, A., Schlegel, C., Granelli, F., Zhang, Y. (eds.) MACOM 2011. LNCS, vol. 6886, pp. 181–192. Springer, Heidelberg (2011). doi:10.1007/978-3-642-23795-9_16

17. Litjens, R., Toh, Y., Zhang, H., Blume, O.: Assessment of the energy efficiency enhancement of future mobile networks. In: Proceedings of WCNC 2014, Istanbul, Turkey (2014)

18. Motorola, 3GPP contribution R1-081638, TBS and MCS Signaling and Tables (2008)

19. 3GPP TS 36.213, Evolved Universal Terrestrial Radio Access (E-UTRA); Physical layer procedures (Table 8.6.1-1 and Table 7.1.7.2-1)

20. FP7 SUNSEED project Deliverable D3.1: Traffic modelling, communication requirements and candidate network solutions for real-time smart grid control, April 2015. http://sunseed-fp7.eu/deliverables/

21. GPP TR 36.912 version 12.0.0, Feasibility study for Further Advancements for E-UTRA (LTE-Advanced), September 2014

22. Meeting report, 3GPP RAN1 NB_IoT Ad Hoc meeting, January 2016

Big Data Processing to Detect Abnormal Behavior in Smart Grids

Béla Genge, Piroska Haller, and István Kiss[✉]

Petru Maior University of Tg. Mures,
N. Iorga, No. 1, Tg. Mures, Mures 540088, Romania
bela.genge@ing.upm.ro, phaller@upm.ro, istvan.kiss@stud.upm.ro

Abstract. This paper proposes a methodology to effectively detect abnormal behavior in Smart Grids. The approach uses a cyber attack impact assessment technique to rank different assets, a cross-association decomposition technique for grouping assets and ultimately to reduce the number of monitored parameters, and an anomaly detection system based on the Gaussian clustering technique. The developed methodology is evaluated in the context of the IEEE 14-bus electricity grid model and three distinct classes of cyber attacks: bus fault attacks, line breaker attacks, and integrity attacks.

Keywords: Big data · Smart grid · Power grid · Anomaly detection · Clustering

1 Introduction

The adoption of modern Information and Communication Technologies (ICT) in the various dimensions of the power grid paved the way towards a novel infrastructural paradigm known as the Smart Grid. The Smart Grid is commonly recognized as the next generation power grid with improved operational benefits of control, reliability and safety, and advanced two-way communication. It facilitates the provisioning of new applications including voltage control with high distributed energy resources (DER) penetration, photovoltaic generation and storage control, load reduction programs, and electric vehicles.

Despite the indisputable advantages, this technological advancement raises significant issues concerning the reliability and security of the communications and control systems the core infrastructure of Smart Grids [1–3]. A significant body of research has been allocated to understanding their possible vulnerabilities [4] and deploying effective attack detection systems [5–8]. Nonetheless, the cyber attack detection techniques need to account for the complexity of the Smart Grid, including the wide range of devices, the various parameters that may be monitored [9], the complexity of information collection and processing. Consequently, this paper embraces the complexity of this problem by developing a methodology that includes: (i) a cyber attack impact assessment technique for assessing the sensitivity of Smart Grid parameters and for ranking assets; (ii) a cross-association decomposition technique for grouping assets sensitive to the

© ICST Institute for Computer Sciences, Social Informatics and Telecommunications Engineering 2017
J. Hu et al. (Eds.): SmartGIFT 2016, LNICST 175, pp. 214–221, 2017.
DOI: 10.1007/978-3-319-47729-9_22

same group of cyber attacks, and ultimately for reducing the number of monitored parameters; and (iii) an anomaly detection system based on the Gaussian clustering technique. The developed methodology is evaluated in the context of the IEEE 14-bus electricity grid model and three distinct classes of cyber attacks.

The remaining of this paper is structured as follows. Section 2 provides an overview of related studies, and Sect. 3 presents the proposed methodology. Experimental results are given in Sect. 4. The paper concludes in Sect. 5.

2 Related Work

We first mention the work of Horkan [16], which showed the various opportunities and design decisions that need to be taken into account for distributing detection devices across an industrial communication network. Berthier et al. [9] identified the parameters that need to be monitored in Advanced Metering Infrastructures for detecting cyber attacks. Berthier et al. [9] also identified the possible location of detection elements, which may be deployed as stand-alone hardware, or may be integrated into smart meters. Zhang et al. [17] developed a detection system provisioned across different layers of the smart grid. Detection modules adopted various techniques including Support Vector Machine and Artificial Immune Systems in order to detect and to classify malicious data and possibly, cyber attacks. Then, Levorato and Mitra [18] elaborated a sparse approximation-based anomaly detector, which perform estimation according to a stochastic model reducing the required amount of observations in a smart grid. Although this method is useful for reduction of monitored signals, introduces issues of how accurate the predicted values represent reality and concerns if sequenced attacks could circumvent this type of detection. Lastly, we mention the work of Cardenas et al. [19], which is the closest to the methodology proposed in this paper. Cardenas et al. [19] developed a cost model-based framework to aid utilities in the provisioning of intrusion detection systems. Compared to [19], this paper adopts big data processing techniques to reduce the set of monitored parameters, and a clustering-technique to detect abnormal behavior.

3 Proposed Methodology

3.1 Requirements Analysis and Overview

According to [10] intrusion detection systems may monitor a wide range of parameters. Berthier et al. [9] identified the main parameters that need to be monitored in Advanced Metering Infrastructures for detecting cyber attacks. Besides the aforementioned aspects, the deployment of detection infrastructures in the Smart Grid must take into account its physical dimension, that is, the underlying physical process. Therefore, we believe that it is imperative to create a solution that tackles the complexity of detecting attacks in the Smart Grid. The methodology presented in this paper (see Fig. 1) ranks Smart Grid assets by embracing the output of a risk assessment methodology, i.e., the cyber attack

Fig. 1. Architecture of the proposed methodology

impact assessment (CAIA). The output of CAIA is then used to group parameters according to their sensitivity to different attacks. This step builds on the cross-association decomposition technique as proposed by Chakrabarti *et al.* in [11], and aims at reducing the number of monitored parameters to those that are the most representative to a given set of attacks. Lastly, the identified parameters are monitored by means of a Gaussian mixture clustering technique and a cluster assessment technique known as *silhouette*. The clustering technique classifies various operational states, e.g., steady states, but it also signals abnormal behavior, which might be attributed to cyber attacks.

3.2   Cyber Attack Impact Assessment Technique

The CAIA technique [12] relies on the following definitions of finite sets: $T = \{1, 2, ..., t, ...m\}$ is the set of measurements for each discrete time moment t, $J = \{1, 2, ..., j, ...n\}$ is the set of observed variables, and $I = \{1, 2, ..., i, ...k\}$ is the set of control variables. Furthermore, we define $Y^i, i \in I$ as a bi-dimensional $m \times n$ matrix containing m measurements of n observed variables for an intervention applied on control variable u_i. We use Y_{tj}^i and Y_{tj}^0 to denote the t-th measurement of the j-th observed variable for a scenario implementing a specific intervention and for a scenario without interventions, respectively. Essentially, CAIA compares the values of observed variables before and after the execution of a specific intervention by means of calculating the cross co-variance between measurement time series. More formally, the following equation defines the mean value of the j-th observed variable for interventions on the i-th control variable:

$$\bar{Y}_j^i = \frac{1}{m} \sum_{t \in T} Y_{tj}^i, \forall i \in I, \forall j \in J \tag{1}$$

Next, we define the intervention-free mean value for the j-th observed variable:

$$\bar{Y}_j^0 = \frac{1}{m} \sum_{t \in T} Y_{tj}^0, \forall j \in J \tag{2}$$

Then, CAIA calculates the cross co-variance between the j-th intervention-free observed variable and the j-th observed variable with intervention on the i-th control variable.

$$C_{ji} = \frac{1}{m} \sum_{t \in T} \frac{(Y_{tj}^i - \bar{Y}_j^i)(Y_{tj}^0 - \bar{Y}_j^0)}{\bar{Y}_j^0 \bar{Y}_j^i}, \forall j \in J, \forall i \in I \tag{3}$$

The values of $C_{ji}, \forall j \in J, \forall i \in I$ are the elements of the C matrix of cross co-variances, which is a detailed result of impact assessment. Accordingly, in [13] CAIA is practically applied in the vulnerability testing of Power Grids.

3.3   Cross-Association Decomposition Technique

The decomposition technique finds the subsets of the sparse binary matrices with the same pattern. The input to this procedure is a binary impact matrix C', com-puted from C such that $C'_{ji} = 1$ if $C_{ji} \geq \alpha$, and $C'_{ji} = 0$, otherwise. α is a thresh-old that is set according to the desired sensitivity level. The Cross-association [11] method seeks the groups of observed variables with similar behavior on a group of attacks regarding the control variables and it automatically determines homogeneous matrix regions and the number of row and column groups. The subset assignments provide an information-theoretic measure of optimality based on the Minimum Description Length principle [21]. As a result, every region will have the same density and this regularity is used to compress the description and to reduce the number of bits required to describe its structure.

3.4   Gaussian Clustering for Anomaly Detection

Based on the output of the previous step, the anomaly detection selects pairs of components from each group of observable variables that exhibit a similar level of sensitivity to the same set of cyber attacks. The approach embraces the Gaussian mixture clustering technique to detect anomalies in Smart Grids. Gaussian mix-ture models are based on combining multivariate normal density components. The clustering process is carried out by an Expectation Maximization (EM) algorithm to select the component that maximizes the posterior probability. In other words, the clustering algorithm uses EM to calculate the probability for an observation to fit to the correct cluster. Since the approach is based on a density measure, it is suitable to identify outlier observations, which in general have very low fit values and may be caused by cyber attacks. The last step of the clustering is the partitioning of the observation vectors into the desired k clusters. By knowing the posterior probabilities, each data point is added to the cluster that gives the largest posterior probability value.

The clustering procedure is followed by the interpretation of results. For this purpose we adopt the advantages of a common internal clustering evaluation technique, known as *silhouette*. According to [14] "The silhouette value for each point is a measure of how similar that point is to points in its own cluster com-pared to points in other clusters, and ranges from −1 to +1". The silhouette evaluation is based on two dissimilarity measures between observations. Addi-tional information on the adopted anomaly detection may be found in [15].

4   Experimental Results

In the evaluation of the proposed methodology we adopted the IEEE 14-bus model. This includes 5 generators, 11 loads, three transformers, 14 buses, and

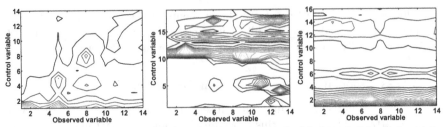

(a) Attack 1 impact matrix. (b) Attack 2 impact matrix. (c) Attack 3 impact matrix.

Fig. 2. The impact matrix of cyber attacks.

20 branches. Besides these, the traditional IEEE 14-bus model was enriched with control loops specific to real-world power systems such as Power System Stabilizer (PSS), Automatic Voltage Regulators (AVR), and Turbine Governors (TG). We further assume three attack scenarios: *Attack 1* triggers bus faults to substation lines by changing control variables in charge of opening/closing bus line circuit breakers; *Attack 2* opens the circuit breakers located on branches, i.e., the lines connecting two substations (buses); and *Attack 3* is an integrity attack on load control devices and on AVRs. In all three cases we measure the impact on the substation's voltage levels. The attacks have been simulated with Matlab, while the IEEE 14-bus electricity grid model was simulated with the Matlab PSAT toolbox [20].

4.1   Cyber Attack Impact Assessment

We applied the CAIA methodology for each of the three attack types. The results depicted in Fig. 2 show the different sensitivity levels of control variables to each attack type. The results also show that, given the wide variety of attacks, and their possible parameters, in general, the exact identification of the most significant control/observed variables is difficult. Therefore, in the next section the output of the CAIA methodology is processed with the cross-association decomposition technique in order to reduce the set of monitored parameters.

4.2   Cross-Association Decomposition

By applying the cross-association decomposition we aim at identifying groups of observed variables that exhibit a similar level of sensitivity to a group of cyber attacks. As mentioned in this paper, the transformation of the C matrix needs to account for the level of desired sensitivity, that is, on the value of α. In order to highlight the effect of α on the result of the cross-association decomposition, we performed different tests with $\alpha = 0.05$, $\alpha = 0.1$, and $\alpha = 0.2$. The results depicted in Fig. 3 show in gray-scale color coding the density and, consequently, the capability to detect a specific type of attack. Here, the black color indicates a high densitivity group where most variables are sensitive to all the attacks

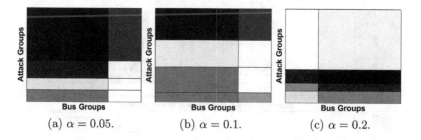

(a) $\alpha = 0.05$. (b) $\alpha = 0.1$. (c) $\alpha = 0.2$.

Fig. 3. The results of the cross-association decomposition.

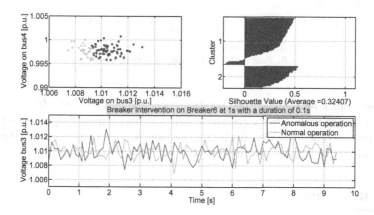

Fig. 4. Non-detected cyber attack.

included in the group. The white color indicates a low densitivity group where variables do not detect the group of attacks. As expected, and confirmed by results (see Fig. 3), the size of the groups in which variables that are not sensitive to a specific group of cyber attacks increases with the value of α. Therefore, a key challenge is to find the optimal selection of α such that all classes of cyber attacks are detected with a minimum number of monitored parameters. However, this procedure needs to account for the sensitivity of the detection technique. These aspects are considered to be part of our future work.

4.3 Clustering-Based Attack Detection

Lastly, we briefly show the applicability of the Gaussian mixture clustering technique to detect cyber attacks on the IEEE 14-bus model. Given the output of the cross-association decomposition, we first select the voltage levels on bus 3 and 4, which belong to the same low-density group. We simulated a 0.1s attack on breaker 6, which, as depicted in Fig. 4, is not detected. However, by selecting two other observable variables from the high-density group, i.e., buses 9 and 12, as depicted in Fig. 5, the attack on the same breaker is successfully detected.

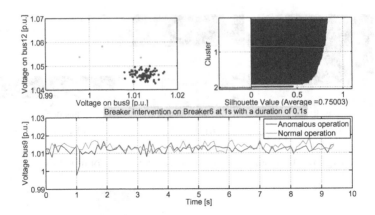

Fig. 5. Successful detection of a cyber attack.

These results have shown the need to correlate and to link different assessment techniques with cyber attack detection methodologies. As proposed in this paper, the outcome of CAIA can be complemented by big data processing techniques to identify the key parameters that need to be monitored.

5   Conclusions

We developed a systematic methodology to detect cyber attacks in Smart Grids. The methodology embraces a cyber attack impact assessment technique to rank Smart grid assets, a cross-association decomposition technique to group and to reduce the number of monitored parameters, and a Gaussian mixture clustering technique to detect cyber attacks. The experiments performed on the IEEE 14-bus model have shown the effectiveness of the proposed approach, and they have indicated the need to address the tuning of parameters in the developed methodology, which we consider to be part of our future work.

Acknowledgement. This work was supported by a Marie Curie FP7 Integration Grant within the 7th European Union Framework Programme (Grant no. PCIG14-GA-2013-631128).

References

1. CrySiS Lab. sKyWIper (a.k.a. Flame a.k.a. Flamer): a complex malware for targeted attacks (2012)
2. Genge, B., Graur, F., Haller, P.: Experimental assessment of network design approaches for protecting industrial control systems. IJCIP **11**, 24–38. Elsevier (2015)
3. Cherepanov, A.: BlackEnergy by the SSHBearDoor: attacks against Ukrainian news media and electric industry (2016)

4. Filippini, R., Silva, A.: A modeling framework for the resilience analysis of networked systems-of-systems based on functional dependencies. Reliab. Eng. Syst. Saf. **125**, 82–91 (2014)
5. Caselli, M., Zambon, E., Kargl, F.: Sequence-aware intrusion detection in industrial control systems. In: Proceedings of the 1st ACM Workshop on Cyber-Physical System Security, pp. 13–24 (2015)
6. Goldenberg, N., Wool, A.: Accurate modeling of Modbus/TCP for intrusion detection in SCADA systems. IJCIP **6**(2), 63–75 (2013)
7. Schuster, F., Paul, A., König, H.: Towards learning normality for anomaly detection in industrial control networks. In: Doyen, G., Waldburger, M., Čeleda, P., Sperotto, A., Stiller, B. (eds.) AIMS 2013. LNCS, vol. 7943, pp. 61–72. Springer, Heidelberg (2013). doi:10.1007/978-3-642-38998-6_8
8. Zhao, J., Liu, K., Wang, W., Liu, Y.: Adaptive fuzzy clustering based anomaly data detection in energy system of steel industry. Inf. Sci. **259**, 335–345 (2014)
9. Berthier, R., Sanders, W.H., Khurana, H.: Intrusion detection for advanced metering infrastructures: requirements and architectural directions. In: First IEEE International Conference on Smart Grid Communications, pp. 350–355 (2010)
10. Scarfone, K.A., Mell, P.M.: Guide to intrusion detection and prevention systems (IDPS). (NIST SP) - 800–94. National Institute of Standards and Technology (2007)
11. Chakrabarti, D., Papadimitriou, S., Modha, S.D., Faloutsos, C.: Fully automatic cross-associations. In: Proceedings of the Tenth ACM SIGKDD International Conference on Knowledge Discovery and Data Mining, pp. 79–88 (2004)
12. Genge, B., Kiss, I., Haller, P.: A system dynamics approach for assessing the impact of cyber attacks on critical infrastructures. Int. J. Crit. Infrastruct. Prot. **10**, 3–17 (2015)
13. Kiss, I., Genge, B., Haller, P., Sebestyen, G.: A framework for testing stealthy attacks in energy grids. In: IEEE International Conference on Intelligent Computer Communication and Processing (ICCP), Cluj-Napoca, pp. 553–560 (2015)
14. Kaufman, L., Rousseeuw, P.J.: Finding Groups in Data: An Introduction to Cluster Analysis, vol. 344. Wiley, Hoboken (2009)
15. Kiss, I., Genge, B., Haller, P.: A clustering-based approach to detect cyber attacks in process control systems. In: 2015 IEEE International Conference on Industrial Informatics, Cambridge, UK, pp. 142–148 (2015)
16. Horkan, M.: Challenges for IDS/IPS Deployment in Industrial Control Systems. SANS Institute Reading Room (2015)
17. Zhang, Y., Wang, L., Sun, W., Green, R.C., Alam, M.: Distributed intrusion detection system in a multi-layer network architecture of smart grids. IEEE Trans. Smart Grid **2**(4), 796–808 (2011)
18. Levorato, M., Mitra, U.: Fast anomaly detection in smart grids via sparse approximation theory. In: 2012 IEEE 7th Sensor Array and Multichannel Signal Processing Workshop (SAM), pp. 5–8 (2012)
19. Cardenas, A.A., Berthier, R., Bobba, R.B., Huh, J.H., Jetcheva, J.G., Grochocki, D., Sanders, W.H.: A framework for evaluating intrusion detection architectures in advanced metering infrastructures. IEEE Trans. Smart Grid **5**(2), 906–915 (2014)
20. Milano, F.: An open source power system analysis toolbox. IEEE Trans. Power Syst. **20**(3), 1199–1206 (2005)
21. Grunwald, P.: A Tutorial Introduction to the Minimum Description Length Principle. Advances in Minimum Description Length: Theory and Applications, pp. 23–81. MIT Press, Cambridge (2005)

[Invited Paper] Native IP Connectivity for Sensors and Actuators in Home Area Network

Pang Zhibo[1], Gargi Bag[1], Edith Ngai[2(✉)], and Victor Leung[3]

[1] ABB Corporate Research, Vasteras, Sweden
{pang.zhibo,gargi.bag}@se.abb.com
[2] Department of Information Technology, Uppsala University, Uppsala, Sweden
edith.ngai@it.uu.se
[3] Department of Electrical and Computer Engineering,
University of British Columbia, Vancouver, Canada
vleung@ece.ubc.ca

Abstract. This paper discusses the requirements and trends in the building automation industry and in general shift of focus to provide native IP connectivity in low power devices. It then provides experimental results that validate the readiness of protocols for Internet of things developed by IETF and available in some chipset to meet the requirements from the industry. The evaluation criteria that was selected in this paper is latency, packer delivery rate, coverage and power consumption. It can be seen from the results that maximum 3 hops can be supported in order to achieve a PDR between 80–90 percent for 150 ms deadline.

Keywords: Building automation · IoT · Performance evaluation

1 Introduction

With the need for effective machine to machine communication to accommodate different requirements of varied domains such as Intelligent Grid and Smart Homes, the systems are rapidly evolving to native IP connectivity [1]. For seamless interactions in between different systems, native internet protocol (IP) support is essential for achieving end to end communication without involving complex and expensive gateways for protocol translation at the network layer. This trend is already evident as quite a number of communication standards both emerging and established; migrating towards IP. For example Zigbee Alliance has proposed Smart Energy Profile that is based on IP and other IETF related protocols [2]. For home and building automation, migrating to IP is quite essential as future devices at home are envisioned to be "smart" for maintaining user's comfort, security, safety and energy efficiency.

© ICST Institute for Computer Sciences, Social Informatics and Telecommunications Engineering 2017
J. Hu et al. (Eds.): SmartGIFT 2016, LNICST 175, pp. 222–231, 2017.
DOI: 10.1007/978-3-319-47729-9_23

1.1 Application Context

Existing systems tend to be more and more interconnected in the future as the concept of smart city, smart home and smart grid becomes reality. This implies that information has to be exchanged between the devices that are located in different networks. For example a device located in the home area network (HAN) may need to obtain demand response signal from the utility containing electricity prices, based on which and user's preference the device can switch to an operational mode. Similarly the same device may need additional information such as weather forecast data from another source to optimize its duty cycle. It is thus very likaly that in the near future a home automation (HA) or building automation (BA) device will have the capability to have end to end communication with a number of devices located both in internal and external networks. Having IP has a network layer removes the bottleneck of having multiple gateways thus improving scalability and reliability. While IP itself has been widely adopted by internet users, the adoption of IP for low power devices with low processing power and low memory is standardized by IETF Internet of Things (IoT) including protocols such as 6LoWPAN, RPL and CoAP. To realize this vision of native IP-based wireless HA it is mandatory to establish the maturity of such an IETF IoT protocol stack in terms of reliability, latency, power consumption, and complexity.

1.2 Main Contribution

In this main paper we have proposed and confirmed the feasibility of native IP wireless for HAN given the expectation to the industrial practices. More specifically we have evaluated off the shelf IETF IoT stack to learn its readiness to meet the requirements of home automation networks in terms of reliability, latency, power consumption and coverage. There is existing work that evaluates the IETF IoT protocols itself such as in [3–5], which looks into different aspects of RPL. Some existing work [6, 7] also validates the IETF IoT stack for smart city use case. However evaluation of the stack for BA/HA requirements is currently missing.

2 Trends and Requirements

2.1 Trends in Building Automation

The automation devices used in the HA or BA are low power devices that have limitation of low memory and low processing power. Thus the low power technologies and protocols developed for wireless sensor and actuators to provide the field devices with native IP connectivity such as IETF IoT and Thread is a promising direction of the evolution of communication technologies for BA systems. It can not only ease the interoperability challenges during the system integration of various devices, sub-systems, and value-added services from different suppliers, but also tear down the walls that are hindering the BA industry to benefit from the vast amount of innovations in internet domain which evolves much faster. Leveraging on added benefits of IP several standards in BA domain such as BACnet [8] is moving towards IP. Other examples include KNX, Bluetooth and DECT ULE [9]. Given the standards which already have native IP connectivity such as the IEEE802.11ah Low Power WiFi, Thread Group [10], and the IETF IoT Suite [11], the BA industry has reached a common consensus to enter the native IP era in near future even though the landscape of standardization is till fragmented. Thus the BA industry is demanding an Industry-Friendly and native IP communication architecture which will not only meet the critical technical performances but also take care the business benefits of all the stakeholders in the value chain.

2.2   Benefits of Native IP

IPv6 was designed to accommodate large number of devices since it has a large address space and this feature is a key component in realizing Internet of Things (IoT). By using IP, large number of devices can be connected in terms of interoperability at the network layer which ensures that communication is end to end and thus transport layer or TCP session is not broken at the gateways and security is also maintained throughout the whole leg of the communication. This not only provides interoperability among different types of devices located in different networks but enables scalability.

2.3   Requirements and Challenges Towards Native IP

The existing wired native IP technologies like BACnet/IP and KNXnet/IP has evolved to meet the requirements of future building automation systems by achieving acceptable performances in terms of latency, reliability, and security. But the wireless native IP technologies is yet to reach the same level of performances as the wired native IP technologies. So the wireless native IP technologies must address the following technical challenges eventually [12, 13].

1. Security: Security is major concern when the BA devices are based on wireless IP. The likelihood of a miscreant sending a command to start all the heavy loads (e.g. air conditioners) that leads to an accident is threatening. At same time data needs to be encrypted to address the privacy concern. Supporting the security protocols and mechanisms in the low power devices with low processing power is a challenge.
2. Reliability: The low power BA devices mostly forming lossy networks based on standards such IEEE 802.15.4 are prone to interferences and distortions of radio signals.
3. Latency: The latency is fundamentally limited by the physical layer bandwidth. Compared with the physical layer bandwidth of wired NIP technologies which is typically 100 Mbps, the bandwidth of radios used in the wireless native IP standards is too narrow e.g. up to 250 Kbps for ZigBee IP and IETF IoT Suite, 1 Mbps for Bluetooth LE, and 1.152 Mbps for DEC ULE.
4. Power consumption: Most of the BA devices will operate on batteries that are expected last at least 3 years. Even some of the devices may need to operate without batteries utilizing energy harvesting techniques.

5. Engineering process: The system integrators and users using the BA devices may need easy to use tools for commissioning and configuring of BA devices. Whereas such tools exist for some of the BA technologies such as KNX, it is ultimately lacking for native IP technologies.

3   IETF IoT

3.1   Protocol Overview

The IoT protocol suit is being standardized by IETF and consists of protocol for transmission of IPv6 packets into IEEE 802.15.4 frames (6LoWPAN), routing in low power lossy networks (RPL), and web transfer protocol for constrained devices (CoAP) as described below in more details.

1. IPv6 over Low-Power Wireless Personal Area Networks (6LoWPANs): The 6LoWPAN working group in IETF has proposed a number of specifications that will enable transmission of IPv6 packets within an IEEE 802.15.4 networks utilizing IEEE 802.15.4 frames. The standard provides guidelines for compressing/decompressing IPv6 headers, fragmentation and reassembly of IPv6 packets, neighbor discovery and bootstrapping.
2. Routing over Low Power Lossy Networks (ROLL): The ROLL working group in IETF focuses mainly on routing in low power lossy network such as IEEE 802.15.4, Bluetooth, powerline communications etc. The work group has proposed RPL protocol that can be used for both for multi-point to point and point to multipoint traffic in low power devices.
3. Constrained RESTful Environment (CoRE): The CoRE working group in IETF has proposed Constrained Application Protocol (CoAP): Is a lightweight web transfer protocol that provides application layer connectivity to the low power devices. CoAP can be used to for communication between the devices in constrained and also between devices on the Internet and in the low power network.

3.2   Platform Overview

Currently the availability IETF IoT protocols is limited to few platforms. For the experimental evaluation of the maturity of IETF IoT stack CC2650 and CC2658 platform from Texas instrument (TI) is selected in this work since they were not only readily available but also offered multi-standard support and low current consumption during transmission and reception, especially CC2650. The detail characteristics of the platforms are given below.

- CC2658: Based on IEEE 802.15.4, CC 2658 supports IETF IoT protocols such as Contiki, IPv6, UDP, 6LoWPAN, COAP and DTLS. Beside it also supports Zigbee application profiles. The current consumption at reception at active-Mode is 20 mA whereas in transmission at 0 dBm it is 24 mA.

- CC2650: CC2650 is a multi-standard platform that supports multiple lower layer protocols such as Bluetooth Smart, Sub-1 GHz and IEEE 802.15.4. Like CC2538 it supports the IETF related protocols such as Contiki, IPv6, UDP, 6LoWPAN, COAP, and DTLS. It consumes very little power; 6.1 mA during transmission in active-mode at 0 dBm and 5.9 mA during reception.

4   Performance Evaluation

The tests performed on CC2650 are broadly classified in two major categories. The first one is coverage test which measures the maximum distance that can be achieved for CC2650 and CC2538 platforms for different transmission power and radios. The second test evaluates the round trip time (RTT) and packet delivery rate (PDR) for different number of hops and traffic loads. Lastly the tests were conducted to measure the current consumption of the CC2650 chipset when performing the performing CoAP operations such as sending a CoAP GET and receiving CoAP ACK from the receiver.

4.1   Communication Range and Reliability

A coverage test is performed to test the coverage of both the Bluetooth Low Energy (BLE) and IEEE 802.15.4 radios in CC2650 chipsets. The four test cases that were used to test the coverage are (a) CC2650 (Bluetooth Low Energy) BLE module is used to send beacons and Google Nexus 5 was used to receive beacon (b) CC2650 BLE module is used both as a transmitter and receiver (c) CC2650 802.15.4 module is used both as a transmitter and a receiver (d) CC2538 border router (BR) acts as CoAP client and CC2538 nodes acts as CoAP severs [14].

These test cases are repeated under different transmission power setting of 3dBm, −3dBm and −9dBm and when sender and the receiver are placed on the same and different floors of the same building. The maximum transmission point in this case is defined as the place where the round-trip time start to increase to abnormal value. Normally, the RTT is 7 ms while the value increases significantly (e.g. to 400 ms + or even pack lost) when the quality of communication is bad. Figure 1 shows the coverage when using different radios with different transmission power with the BR and the test device located in the same floor. This test was repeated with test nodes placed in floors from the BR.

Table 1 provides the detail result of the coverage test. It can be seen from Table 1 that when the sender and receiver is placed in the same floor the distance covered is longest for 3 dBm and the shortest for −3 dBm. When the sender is placed one floor below the receiver it can be seen that coverage in the horizontal distance is getting short as some power is used to cover the vertical distance between the sender and the receiver. When the nodes are placed in the basement communication is only possible using 3 dBm. Also it can be observed that single hop has a distance of around 60 m in the same floor for 3 dBm transmission power.

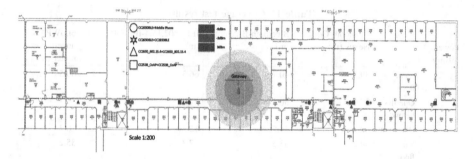

Fig. 1. Range of different radios under different transmission power when nodes are placed in same floor as BR

4.2   Latency

Figure 2 shows an example of the topology that is used to determine the PDR and the RTT of the CC2650 devices with no duty cycling. More specifically 20 nodes were used including 1 BR. The size of the BR CoAP GET message used was 53 bytes, whereas it was 63 bytes for CoAP ACK. The nodes were deployed in a typical office facility and building had sources of interference from other networks such as WiFI, Bluetooth, Wireless HART and other electromagnetic interference from high voltage laboratories. In general the following test cases were considered [15].

- **When there is 2 transmission/s between BR and a node and load in the network is Zero:** In Fig. 2, the BR acts as a CoAP client and the other nodes in the network acts a server. Every 0.5 s the router sends CoAP GET message to a selected node and waits for ACK message. The total transaction for a node is 300. There is no extra traffic in the network.
- **Burst Traffic between BR and node and load in the network is 0.2/transmission/s/node:** The router sends a test packet (CoAP GET) immediately after either receives a response or there is a time out. The total transaction for a node is 300. Meanwhile every node in the network sends CoAP GET message to a random node and waits for ACK message with time interval of 5 s.
- **Burst traffic between BR and node and load in the network is zero:** The router sends a test packet (CoAP GET) immediately it either receives a response or there is a time out. The total transaction for a node is 300. There is no extra traffic in the network.
- **For 2 transmission/s between BR and node and load in the network is 0.2/transmission/s/node:** Every 0.5 s the router sends CoAP GET message to a selected node and waits for ACK message. The total transaction for a node is 300. Meanwhile every node in the network sends CoAP GET message to a random node and waits for ACK message with time interval of 5 s.

Figure 3 shows the packet delivery rate of the above four cases varying hop count. By comparing the PDR at 150 ms it can be seen that 80 to 90 percent of the packets arrive within 150 ms deadline for all the four load cases for 3 hops. However for more number of hops the PDR is degrading especially when there is load in the network.

Table 1. Coverage test results

Tx Power (dBm)/	Floor	Bluetooth-Google Nexus (m)	Bluetooth-Bluetooth (m)	IEEE802.15.4-IEEE802.15.4 (m)	6LoWPAN node-border router (m)
3	Same floor	39.5	40.7	66.3	60.5
	One floor below	27.6	30.5	34.0	28.7
	Basement	6.8	11.5	12.3	10.3
−3	Same floor	34.2	33.5	37.3	35.2
	One floor below	19.5	21.0	24.5	23.5
	Basement	No	No	No	No
−9	Same floor	21.8	21.1	28.3	28.3
	One floor below	11.4	12.4	16.3	14.1
	Basement	No	No	No	No

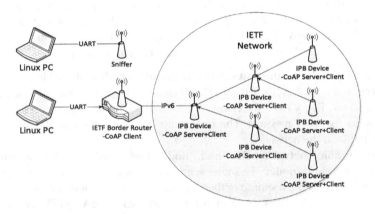

Fig. 2. An example of the network topology

Similarly it can be also seen that PDR is increasing for increasing deadline. For example for 1000 ms deadline the PDR is above 90 percent for all the four load cases especially for 3 hops. This is because the more amount of time a node waits, the percentage of packets received is more.

Figure 4 shows the Cumulative Distribution Function (CDF) of the RTT for different number of hops. The brown vertical lines on the graph shown the three deadlines (150 ms, 250 ms, 1000 ms). It can be seen from Fig. 5 that increasing number of hops has a considerable negative effect on the RTT. For example for 1 hop 90 to 92 percent of the packets (depending on the cases) arrive within the 150 ms deadline whereas the value drops to 42 to 89 percent for 6 hops.

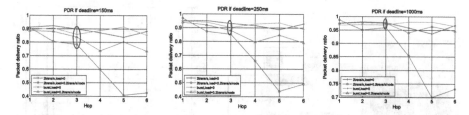

Fig. 3. Packet delivery rate versus hops for (a) Deadline 150 ms (b) Deadline 250 ms (c) Deadline 1000 ms

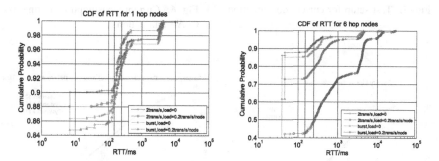

Fig. 4. Cumulative distributed frequency for RTT depending on number of hops

4.3 Power Consumption

In Fig. 5, the current consumption of the CC2650 leaf nodes were measured for radio duty cycling (RDC). The tests were done to evaluate the current consumption of both when the node is behaving as a CoAP client and CoAP server. The packet size of CoAP GET used in the experiment is 63 bytes and CoAP ACK had a size of 83 bytes [12].

The measurements such as current for transmission of CoAP request, current for housekeeping (both for network and device level), the current consumption for post and pre-processing were obtained from Fig. 6 and were used to derive the battery life using the following relationships.

- Current for useful operations (I_{opt})
  - Operation level:
    Operation = $\sum$ Transactions
  - Transaction level:
    Transaction = $\sum$ Reception + $\sum$ Transmission
  - State level:
    Reception = $\sum$ Processing + $\sum$ Radio-Rx
    Transmission = $\sum$ Processing + $\sum$ Radio-CCA + $\sum$ Radio-Tx

From BA perspective useful operations can be for example using button to switch lamps, draw curtains, and control the HVAC etc.

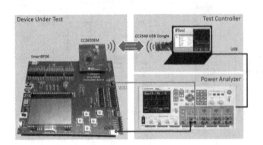

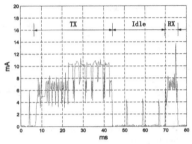

Fig. 5. Test setup for current consumption

Fig. 6. Current consumption during one CoAP transaction

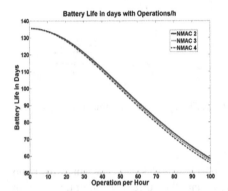

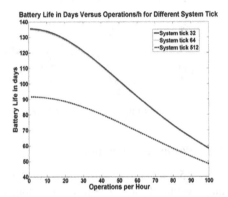

Fig. 7. Battery life in days

The battery life for CC2650 device is given with respect to number of MAC level retries the device makes to transmit and also with respect to the system tick (device level housekeeping). Figure 7 shows that as operations per hour increases and having a large system tick give considerable negative effect on the battery life.

5 Conclusion

As systems are getting more and more interconnected, IP based IETF IoT is going to revolutionize communication since it aims to bring interoperability among devices and connect low power devices to a larger network. In this work CC2650 platform supporting IETF IoT is evaluated to check IETF IoT's readiness to meet BA industry's requirements. It can be seen that PDR for 150 ms deadline varies between 80 to 90 percent (depending on the load in the network) for 3 hops. After 3 hops the performance is degrading rapidly especially when there is load in the network. Also from the coverage test it can be seen that single hop has a distance of around 60 m in the same floor for 3 dBm transmission power and thus 2–3 hops for the 3-building office complex (typical medium size). Moreover depending on the number of operations, the frequency of housekeeping and channel conditions, the maximum battery life that can be attained is around 137 days since even if there is no operations battery will get depleted through discharge current.

References

1. Branger, J., Pang, Z.: From automated home to integrated sustainable, healthy and manufacturing homes: a new story enabled by the Internet-of-Things and industry 4.0. J. Manage. Analytics. Taylor & Francis. doi:10.1080/23270012.2015.1115379
2. Pang, Z., Cheng, Y., Johansson, M.E., Bag, G.: Preliminary study on industry-friendly and native-IP wireless communications for building automation. In: International Conference on Industrial Networks and Intelligent Systems (INISCom2015), March 2015
3. Banh, M., Mac, H., Nguyen, N., Phung, K., Thanh, N., Steenhaut, K.: Performance evaluation of multiple RPL routing tree instances for Internet of Things applications. In: IEEE ATC, Ho Chi Minh City (2015)
4. Accettura, N., Camacho, C., Grieco, L., Boggia, G., Camarda, P.: Performance assessment and tuning rules for low-power and lossy stacks. In: IEEE WFCS, Lemgo (2012)
5. Qasem, M., Altawssi, H., Yassien, M., Dubai, A.: Performance evaluation of RPL objective functions. In: IEEE CIT/IUCC/DASC/PICOM, Liverpool (2015)
6. Isern, J., Betzler, A., Gomezy, C., Demirkoly, I., Paradells, J.: Large-scale performance evaluation of the IETF Internet of Things protocol suite for smart-city solutions. In: ACM WASUN, Cancun (2015)
7. Paventhanyz, A., Darshini, D., Krishnaz, H., Pahujax, N., Khan, M., Jain, A.: Experimental evaluation of IETF 6TiSCH in the context of smart grid. In: IEEE WF-IoT, Milan (2015)
8. BACnet Official Website. www.bacnet.org. Accessed 21 Oct 2014
9. IETF 6Lo Working Group, Transmission of IPv6 Packets over DECT Ultra Low Energy (draft-mariager-6lowpan-v6over-dect-ule-03), 15 July 2013
10. The Thread Group. http://threadgroup.org. Accessed 21 Oct 2014
11. Palattella, M.R., Accettura, N., Vilajosana, X., Watteyne, T., Grieco, L.A., Boggia, G., Dohler, M.: Standardized protocol stack for the Internet of (important) Things. IEEE Commun. Surv. Tutorials **15**(3) (2013). doi:10.1109/SURV.2012.111412.00158
12. Wang, J., Pang, Z., Pang, C., Vyatkin, V.: Industry-friendly engineering tools for wireless home automation devices. In: INDIN 2015 IEEE International Conference on Industrial Informatics, Cambridge, UK
13. Wang, J., Pang, Z., Bag, G., Johansson, M.E.: RESTful information exchange among engineering tools for wireless home automation devices. In: The 2015 International Conference on Computer, Information, and Telecommunication Systems, CITS 2015, Gijon, Spain (2015)
14. Bag, G., Pang, Z., Johansson, M., Min, X., Zhu, S.: Wireless building automation devices supporting multiple standards: challenges and feasibility. In: IEEE ICIEA, China (2016, to appear)
15. Bag, G., Pang, Z., Johansson, M., Min, X., Zhu, S.: Engineering friendly tool to estimate battery life of a wireless sensor node. In: IEEE WFCS, Portugal (2016, to appear)

Author Index

Printed in the United States
By Bookmasters